Analytical Chemistry of Foods

Analytical Chemistry of Foods

Edited by:

Sujata Nagnath Mustapure

ARCLER PRESS

www.arclerpress.com

Analytical Chemistry of Foods

Sujata Nagnath Mustapure

Arcler Press
224 Shoreacres Road
Burlington, ON L7L 2H2
Canada
www.arclerpress.com
Email: orders@arclereducation.com

ISBN: 978-1-77407-625-5 (Hardcover)

ABOUT THE EDITOR

Dr. Sujata Nagnath Mustapure is an associate professor at Department of Electrical & Other Energy sources, College of agricultural engineering & Technology, Vasantrao Naik Marathwada Krishi Vidhyapeet, Parbhani. She obtained her B.Tech (agril. Engg.) in 2010 from Vasantrao Naik Marathwada Krishi Vidhyapeet, Parbhani. Gold medal in M.Tech (Agril. Engg.), 2014 from Dr. Panjabrao Deshmukh Krishi Vidyapeeth, Akola and Ph.D in Renewable energy engineering in 2019 from Maharana Pratap University of Agriculture and Technology, Udaipur. She has also worked as assistant professor in College of Agricultural, Naigoan, Vasantrao Naik Marathwada Krishi Vidhyapeet, Parbhani. She has authored several national and international publications.

TABLE OF CONTENTS

LIST OF FIGURES

LIST OF TABLES

LIST OF ABBREVIATIONS

AAA	Amino acid analysis
AQUA	Assays utilizing absolute quantification
ARP	Amadori rearrangement product
AUC	Analytical ultracentrifugation
BET	Brunauer-Emmert-teller
BHA	Butylated hydroxyanisole
BHT	Butylated hydroxytoluene
C	Carbon
CD	Circular dichroism
CE	Capillary electrophoresis
CI	Chemical ionization
Cl	Chloride
COOH	Carboxyl group
DDME	Drop-to-drop microextraction
DI	Direct-immersion
DLLME	Dispersive liquid-liquid microextraction
DNA	Deoxyribonucleic acid
DSC	Differential scanning calorimetry
DSPandUSP	Down-stream and up-stream processing
DSPE	Dispersive solid-phase extraction
EI	Electron impact ionization
ELISA	Enzyme-linked immunosorbent assay
ELSD	Evaporative-light scattering detectors
ERH	Equilibrium relative humidity
ESI	Electrospray interface
ESR	Electron spin resonance

ESY	Extracting syringe
FAMEs	Fatty acid methyl esters
FFF	Field-flow fractionation
FID	Flame ionization detector
FOS	Fructooligosaccharides
FTIR	Fourier transforms infrared spectrometry
GC	Gas chromatography
GCB	Graphitized carbon black
GC-MS	Gas chromatography-mass spectrometry
GOS	Galacto oligosaccharides
H_2O	Water
HF(2)LPME	Hollow fiber-protected two-phase liquid mi-cryoextraction
HILIC	Hydrophilic interaction liquid chromatography
HMF	Hydroxymethylfurfural
HMF^{one}	4-hydroxy-5-methyl-2, 3-dihydrofuran-3-one
HOS	Higher-order structure
HPAEC	High-performance anion-exchange chromatography
HPLC	High-pressure liquid chromatography
HS	Head-space
ICP-AES	Inductively attached plasma-atomic emission spectrometry
IEC	Ion exchange chromatography
IMF	Intermediate moisture foods
IR	Infrared
IRMS	Isotope ratio mass spectrometry
IUB-MB	International union of biochemistry and molecular biology
LDH	Lactate dehydrogenase
LC	Liquid chromatography
LLE	Liquid-liquid extraction
LLLME	Liquid-liquid-liquid microextraction
LOD	Limit of detection
LPME	Liquid phase microextraction
MAE	Microwave-assisted extraction
MALDI-TOF	Matrix-assisted laser desorption/ionization-time of flight
MIPs	Molecular imprinted polymers

MISPE	Molecular imprinted solid-phase extraction
MLipid	Lipid is measured
MOS	Mannan oligosaccharides
MRLs	maximum residue levels
MS	Mass spectrometry
M_{Sample}	Mass of initial sample
MSPD	Matrix solid-phase dispersion
MW	Molecular weight
Na+	Sodium
NH_2	Amino group
NMR	Nuclear magnetic resonance
NP-HPLC	Normal phase HPLC
O_2	Oxygen
OH	Hydroxyl
PandT	Purge-and-trap
PAD	Pulsed amperometry detection
PCA	Principal component analysis
PDI	Plasma desorption ionization
pI	Isoelectric point
PLE	Pressurized liquid extraction
PPS	Protagen protein services
PQQ	Pyrroloquinoline quinone
PSA	Primary secondary amine
PTV	Programmed thermal vaporization
RAM	Restricted access media
RI	Refractive index
RS	Raman spectroscopy
RSD	Relative standard deviation
SBSE	Stir bar sorptive extraction
SDE	Simultaneous distillation-extraction
SDME	Single drop microextraction
SEC	Size exclusion chromatography
SFC	Solid fat content
SFE	Supercritical fluid extraction

SMBC	Simulated moving bed chromatography
SPE	Solid-phase extraction
SPME	Solid-phase microextraction
SSC	Single short column
SV-AUC	Sedimentation velocity analytical ultracentrifugation
SWCNs	Single-walled carbon nanotubes
SWE	SSubcritical water extraction
TBA	Thiobarbituric acid
TBARS	Thiobarbituric acid reactive substances
TBHQ	Tertiary-butyl hydroquinone
TLC	Thin-layer chromatography
TMAMQ	Tetramethoxy azobismethylene quinone
Trp	Tryptophan
Tyr	Tyrosine
USE	Ultrasound-assisted extraction
UV	Ultraviolet
Vis	Visible
Wg	Water content
WHC	Water holding capacity

PREFACE

Food chemistry deals with the studies of chemical interactions and processes of all food constituents (both non-biological and biological). The biological food components include beer, meat, milk poultry, and lettuce. Food chemistry has resemblance with biochemistry due to the investigation of constituents such as lipids, proteins, and carbohydrates. However, it also includes discussions of other components such as vitamins, water, minerals, food additives, enzymes, food colors, and food flavors. This field of science also covers the modification in foods caused by different processing techniques. Food conversion phenomena include the fermentation of dairy foods with the help of microorganisms that transform lactose to a new fermented product (i.e., lactic acid).

Analytical Chemistry of foods is a science which deals with attaining, treating, and sharing data about structure and composition of matter. Food quality and safety are also investigated using analytical techniques. Food Analytical techniques mainly encompass the specific and fundamental characteristics of optimization, development, and validation stages of food analysis.

The world has witnessed rapid advances in the field of food technology and food chemistry in the past twenty years. These stupendous developments in the food industry can be attributed to advanced manufacturing and analytical technologies. However, the consolidated data on analytical food chemistry is not widely available for the readers. This book is intended to fill this void.

There are eight chapters in this book. All the chapters are organized in the order of progressive knowledge regarding food chemistry. The first chapter of the book introduces the readers with the fundamentals of food chemistry. History of food chemistry and different types of food constituents are introduced in the chapter. Chapter 2 mainly focuses on a detailed description of the key food components including water, lipids, proteins, carbohydrates, and enzymes. Chapter 3 discusses the sampling procedures and sample preparation techniques used in food analysis.

Chapter 4 discusses the fundamental characteristics of water and its role in food technology and food chemistry. Chapter 5 focuses on the fundamental properties of proteins and their characterization using analytical techniques.

Chapter 6 introduces the readers with different types of carbohydrates. Different characterization methods used for analyzing carbohydrates are also discussed in the chapter.

Basic features and characteristics of lipids are discussed in Chapter 7. Chemical analysis of lipid compounds is also illustrated in the chapter. Finally, Chapter 8 introduces the readers with the fundamentals of enzymes and their use in food compounds. Moreover, Chapter 8 also focuses on the characterization of enzymes using different techniques.

This book is mainly written for a general audience. However, it can be regarded as a ready reference for the students, researchers, scientists, and professors of food chemistry. Moreover over, this book can be useful for readers of multidisciplinary subjects. Businessmen from the food industry can also benefit from the book on analytical food science.

Chapter 1

Fundamentals of Food Chemistry

CONTENTS

1.1 INTRODUCTION

Food science is a multidisciplinary topic including biology, chemistry, engineering, and microbiology food chemistry is a science that deals with properties and the chemical composition of food and chemical variations it undergoes (Maga, 1982; Belitz et al., 2004; Otles, 2005).

The field of food chemistry as an autonomous branch of the science was formed in the nineteenth century in parallel with the increasing interest in the quality of food and the overturning of food falsification and adulteration (Mottram et al., 2002; Stadler et al., 2002). New techniques of food analysis permitted us to characterize and discover not only the main nutrients (such as lipids, proteins, and carbohydrates) but also important minor components like vitamins and important minerals and lastly, thousands of micro and minor constituents being natural components of food (such as natural antioxidants, natural coloring matters, aroma compounds,). The results declared above created the basis of a precise evaluation of the nutritious value of foods, the detection of falsification and adulteration of foods and the determination of quality. The latest activity of food chemistry is mainly connected with the safety of food with a supply of healthy food to the population. A significant part of the activity is the determination and detection of food toxins and control of the food additives (Calvert, 1985; Morton and MacLeod, 1986; Cui, 2005).The food chemistry is a science that deals with the chemical composition of the foods, with properties and chemical structure of food components and with the chemical variations food undergoes during storage and processing (Fennema, 1985; Dombrowski et al., 2005). The food chemist depends heavily on the knowledge of chemistry in his work. Conversely, food chemistry is also associated with microbiology (food safety, preservation, and spoilage), the science of nutrition (nutritious value of foods, pollutants) and biochemistry (especially in the study of the chemical changes of food components). The material comprised in theme food chemistry is separated into chapters according to the main groups of food components: enzymes and proteins, vitamins, flavors, carbohydrates, lipids, and colorants, other micro components and minerals, contaminants, and additives (Hudson, 1981). Only certain tables providing information regarding the gross chemical composition of the core types of foods (milk, eggs, legumes, fruits, oilseeds, bread, cereals, bread, and meat) are comprised in this subject level chapter.

The main sources in each group of compounds, food safety aspects and nutritional requirements are concisely treated. At the start, a short historical outline is also given (Weidenborner, 2001; Sikorski and Kolakowska, 2004).

1.2. HISTORICAL ASPECTS

Even though food science as a distinct science was developed in the second half of the nineteenth century, the acceptable food supply has usually been a central issue for governments, as the creation of ancient societies many thousand years ago. Food laws were generally among the earliest identified by humans. The protection of consumers in the field of falsification and adulteration of food signifies one of the most primitive forms of government controlling of the commercial enterprises. Food regulations prevailed in Egyptian, Chinese, Hindu, Hebrew, Greek, Arab societies, Roman (Schmidt et al., 2005; Risch, 2009).

Trade Guilds were made in the middle ages whose aim was to deliver control and general regulation over the integrity and honesty of the members and quality of the products. The activity of making Trade Guilds sustained till the eighteenth century when the revolution of 1789 in France took away the authority and hierarchies of Guild, and freedom of trade and industry was proclaimed (Vrabcheva et al., 2000).

The eighteenth and the nineteenth centuries were an era of big variations in society and economy but also in the field of science. The advancements in chemistry in this particular era laid the basis of food chemistry as a distinct branch of science. The efforts of French chemist Lavoisier added to the final elimination of the phlogiston principle. Lavoisier established the basic principles of organic analysis and combustion and made the first efforts to conclude the elemental composition of alcohol and stated on organic acids of several fruits. A Swedish pharmacist Scheele discovered glycerol, chlorine, isolated malic and citric acid from various fruits and tested 20 common fruits for malic-citric- and tartaric acids. Liebig besides the discovery of newer food components also improved the analytical techniques, and first categorized the foods as either non-nitrogenous or nitrogenous. Even though this classification wasn't accurate in detail and Liebig had no information about the structure of carbohydrates and proteins, and several other food constituents, which Liebig published in 1847, what is seemingly the first book related to food chemistry entitled "*Researches on Chemistry of Food*" (Iwai et al., 2005).

Pasteur's discovery of microorganisms and information on fermentation procedures was an additional step in development. By the end of the nineteenth century, the development of the chemical analytical methods, advancement in nutrition and physiology made information of the major chemical components of foods probable.

The beginning of the period with the industrial uprising was a time of incredible expansion in numerous fields, which had a specific bearing on the food quality control. Change from a rural society to an urbanized society and from domestic to the factory system in the production of food, with the absorption of populations, placed stress on the production and distribution of food. This period created several public health problems too, because of the bad hygienic conditions, falsification, and adulteration of food and low quality of the foods. Such a situation stimulated struggles of governments and bigger cities to develop institutions to control the chemical composition of foods. For instance, a municipal service was developed in Amsterdam in 1858 for control of the food and beverages. Similar institutions appeared in several countries in Northern America and Europe. The regulations about food were reviewed and the first contemporary food laws were endorsed in several countries. The growing significance of food chemistry preferred the establishment of the agricultural experimental stations, research institutions, food control laboratories, foundation of the scientific journals related to food chemistry and training of the professionals in food chemistry (Shukla et al., 2017).

Although in contrast with food quality and food supply in the nineteenth century the existing situation in this specific field, particularly in developed countries, appears to be perfect, the food safety and food quality remained in the emphasis of the activity of the food chemistry. New raw materials, new foods, new technologies, wide utilization of chemical additives, increasing environmental pollution, growing interest in association with food-health, defends the struggles of food chemists. Globalization of the world food market requires the harmonization of food laws and food regulations. Due to this international organizations were developed in this field.

1.3. GROSS CHEMICAL COMPOSITION OF THE FOODS

As an outcome of examinations of the food quality control positions and several other institutions, currently, data banks, several books, and numerous other facilities are accessible for obtaining information regarding the chemical composition of the foods (Gross et al., 1989; Renaud and Parry,

1994; Aletor et al., 2002). Keeping in the notice that the structure of food chemistry is centered on the type of components and not on diverse types of foods, in this chapter only a short evaluation will be given regarding character and gross chemical composition of these foods (Ortega-Calvo et al., 1993; Renaud et al., 1999, 2002).

1.3.1. Cereals and their Products

Cereals are basic foods in several countries and generally play the main role in the supply of nutrients. Chemical composition of the cereals grains is usually categorized by a high amount of starch, a substance with comparatively substantial protein content and comparatively low lipid content (Rasper and Walker, 2000; Ross et al., 2003). From minor ingredients, the B-group vitamins, phosphorus, and potassium and magnesium are the chief ones. The gross chemical composition of the commercially significant cereals is given in Table 1.1.

Table 1.1: Gross Chemical Composition of the Cereal Grains

Cereal grain	Protein	Lipid	Starch	Fiber	Ash
Wheat	12.2	1.9	71.9	1.9	1.7
Triticale	11.9	1.8	71.9	1.9	1.8
Rye	11.6	1.7	71.9	1.9	2.0
Oats	10.2	5.8	55.5	10.9	3.2
Barley	10.9	2.3	73.5	4.3	2.4
Millet	10.3	4.5	58.9	8.7	4.7
Maize	10.2	4.6	79.5	2.3	1.3
Rice (brown)	8.1	2.3	75.8	1.8	1.4
Sorghum	11.0	3.5	65.0	4.9	2.6

The several constituents of grain aren't distributed uniformly in the kernel; thus the ratio of components in diverse botanical parts of kernel varies significantly (Zeleny and Coleman, 1938).

An embryo is quite rich in proteins (20–35%) and lipids (12–35%). Consequently, their content of carbohydrate is lower. Embryo also possesses a higher content of minerals and some vitamins. *Endosperm* comprises the highest content of starch and the content of protein is slightly lower than that of the whole kernel. The endosperm is normally the lowest in vitamin, lipid, and mineral content. The *pericarp's chief* component is cellulose, substantial amounts of pentosans, hemicelluloses, and some other non-starch polysaccharides and some lignin are existent. No starch was discovered in hulls, the lipid and protein content is quite low. Phosphorus and potassium are the major mineral components. Silicon is the major mineral component in the rice husk. Coloring matters and waxes are the remaining constituents. The outermost layer of endosperm is the *aleurone layer*. From a nutritional perspective the high protein (24%), lipid (8%), and mineral (11%) content might be mentioned. The aleuron layer is quite rich in vitamins of the B-group. A substantial component is a phytic acid that is existent in the form of granules comprising calcium and potassium salts of the compound. About 70–75% of phosphorus in the cereals happens as phytic acid and is the chief phosphorus stock of the seed (Fardet et al., 2008).

Because of their kernel structure and low moisture content the cereal grains possess a good storability and a broad range of the cereal-based products are utilized worldwide. The chemical composition of the *cereal products* is dependent primarily on the type of the milling product utilized and naturally on the amount and type of other ingredients.

1.3.2. Legumes and Oilseeds

Oilseeds and Grain legumes are very rich in protein among the food raw materials (Table 1.2). Seeds of the leguminous plants have traditionally been consumed by humans and they create a significant protein complement to starchy foods and cereals (Mosse, 1990; de Almeida Costa et al., 2006).

Table 1.2: Percentage Chemical Composition of Legumes and Oilseeds

Legume (oil-seed)	Moisture	Protein	Lipid	Carbohydrate
Pea	12.0	23.3	2.0	69.2
Peanut	9.1	25.0	49.9	20.0
Field bean	12.0	30.1	2.2	61.0
Rapeseed	5.2	26.0	33.8	34.9
Sunflower seed	6.0	30.0	43.5	26.2
Soybean	7.0	45.2	25.3	16.0

The growing issues linked with the supply of the food protein to the population of the world encouraged the search of some new protein sources for food purposes. The efforts to escalate the production of food protein led to increasing commercial production of the plant protein isolates and concentrates and their utilization as ingredients in several new and traditional food products (Amarteifio and Moholo, 1998; Choct and Hughes, 1999). Particularly the soy protein production concentrates were increased because of its auspicious chemical composition. After the removal of oil under slight conditions the non-protein constituents are eliminated and concentrates and isolates might be obtained (Choct, 1997; Enujiugha, 2003).

1.3.3. Vegetables and Fruits

The terms vegetables and fruits cover a very big number of plants or more accurately edible parts of plants. From the botanical viewpoint there are *roots* like sugar beet, yams, sweet potato, and cassava; *stems* like potato, onion, rhubarb, asparagus, and kohlrabi; *leaf crops* like spinach and lettuce; and *fruits*. The edible parts of the plant are composed of living tissues that are usually metabolically active even after harvesting. It means they are less or more constantly varying in composition dependent on increasing conditions former to harvest, the genetic pool of cell, physiological role in the parts of a plant and the environment after harvesting. *Water* normally represents about 70 to 90% of fresh weight. The main components of a solid matter of the plant tissues are *carbohydrates* (Van Buren, 1979; Saini et al., 2015).

Sugars are the leading carbohydrate component in fruits, whereas starch in potato and the other starchy vegetables. In addition to starch and sugars, digestible constituents of the plant tissues, many compounds namely *cell wall constituents* are existent in vegetables and fruits. The principal cell wall components are cellulose, pectin, hemicelluloses, and lignin. The comparative proportion of these compounds is dependent on the species, degree of ripeness at harvest and environments of storage after the harvesting. Even though the cell wall components, the main components of the dietary fiber, aren't digestible, there is substantial evidence for the advantageous role played by fiber in the health and disease (Andrés-Lacueva et al., 2010).

Table 1.3: Percentage Chemical Composition of Fruits and Vegetables

Fruit/vegetable	Water	Protein	Carbohydrate	Ash
Apple	90.5	0.4	7.0	0.4
Sour cherry	85.7	0.8	11.0	0.6
Apricot	86.1	0.9	10.2	0.7
Banana	74.1	1.3	22.8	0.9
Orange	89.6	0.9	7.0	0.5
Hazelnut	3.0	15.6	8.7	2.0
Walnut	8.1	18.6	11.7	1.7
Potato	73.0	2.6	21.1	0.9
Green pea	75.1	2.5	13.5	1.0
Tomato	94.1	0.8	3.8	0.6
Spinach	91.5	2.5	1.7	1.6
Carrot	90.1	1.1	7.2	0.8
Cucumber	96.8	0.6	1.8	0.6
Lettuce	94.6	1.8	1.2	0.9
Green pepper	89.5	1.1	5.2	0.6

The *protein* content of vegetables and fruits changes in different species, but it generally signifies only a small percentage (%) of the fresh weight and potato doesn't play a substantial role in the protein supply of the population. Plant tissues might contain a substantial amount of non-protein nitrogen like free amines and amino acids (Sundram et al., 2003; Tsao, 2010).

Vegetables and fruits are a vital source of *vitamins* like vitamin-C, niacin, riboflavin, thiamine, and pro-vitamins of vitamin-A. The overall *mineral content* of vegetables and fruits expressed as ash changes in a wide range (0.1–5%). Potassium, calcium, phosphorus, and magnesium are the main minerals.Among other constituents, the organic acids of fruits, pigments, and aroma compounds may be mentioned.

1.3.4. Fish, Meat, and Their Products

Fish, meat, and their products usually play a key role in offering high-class protein to the entire world population. From a quantitative viewpoint, *water* is the main component of fish and meat (Table 1.4). Its proportion is mainly dependent on the fat content (Gray et al., 1994; Mottram, 1998). Generally, it might be calculated that water plus fat content is constant. The protein content of the lean portion of fish and meat is 19–23%.

Table 1.4: Percentage Chemical Composition of the Meat

Type of animal	Protein	Fat	Water
Cattle (lean)	20.6	3.5	74.2
Pig (lean)	20.1	6.3	78.7
Cattle (fatty)	18.9	24.5	58.3
Lamb (lean)	19.0	6.4	72.1
Pig (fatty)	15.1	35.0	49.0
Broiler	19.8	11.5	67.5
Lamb (fatty)	17.6	26.4	53.5
Duck	13.0	26.8	51.0
Turkey	19.9	8.2	68.4
Carp	20.5	6.1	73.3
Goose	12.2	32.1	52.1
Herring	18.6	3.5	78.0
Trout	23.1	8.8	67.2
Codfish	23.9	1.0	76.1

The main part of the protein is known as contractile protein, other groups are called the soluble proteins and the protein of connective tissues like collagen, elastane, and keratin (Andrée et al., 2010). In avians and mammalians, fat is situated partly in adipose tissue and partly in muscle tissue. Chicken muscles usually contain less fat as compared to those of the mammalians. In the lean fish, fat is normally carried in the liver. In the fatty fish, depot fat happens as the extracellular droplets in muscle tissue (Demirezen and Uruç, 2006; Pearson, 2013).

Muscle tissue is a vital source of *vitamins* mainly vitamins of the B-group (thiamin, niacin, riboflavin vitamin B_{12}). Vitamin-C is existent in only low levels in the meat. Meat is a significant source *of phosphorus and iron* a relatively poor source of calcium (Piironen et al., 2002; Benjakul et al., 2005).

1.3.5. Milk and the Dairy Products

The solid stuff of milk, comprising 12 to 15% of fresh milk is an intricate mixture of proteins, vitamins, lipids carbohydrates and minerals (Table 1.5.). Bovine milk comprises 3.0 to 3.5% *high-quality proteins.* The proteins are categorized as whey- and casein proteins. Casein contains the largest fraction. It might be precipitated with rennet or acid in processes of production of cheese and yogurt respectively. The average fat content is existent in milk in the globules of around 2 to 3 μm in diameter (Palmer, 1924; Creamer and MacGibbon, 1996). It is made up of triacylglycerols and to some extent amounts of sterols and phospholipids (0.1 g/l). *Lactose, milk sugar* is generally the major carbohydrate of milk. Lactose intolerance is occasionally witnessed in some individuals. This is linked with a lack of the digestive enzyme, lactase excruciating lactose to the simple sugars. Indications of lactose mal-absorption are bloating, diarrhea, and abdominal cramps. In the acid milk products like kefir and yogurt, the lactose is converted to lactic acid, thus lactose intolerant individuals can ingest such products freely (Jack and Smith, 1956; Huppertz and Kelly, 2006).

Milk is an outstanding source of *water*, *fat-soluble minerals*, and *vitamins*, mainly calcium and iron. The chemical composition of the milk products is displayed in Table.1.5.

Table 1.5: Chemical Composition of Some Dairy Products and Milk

Product	Water	Protein	Fat	Carbohydrate
Cow milk	87.3	3.6	3.8	4.7
Goat milk	87.4	2.9	4.5	4.1
Sheep milk	80.1	5.5	7.4	4.8
Processed cheese	45–58	18–20	20–25	–
Hard cheese	38–48	22–27	28–36	–
Butter	16–20	1	78–82	1
Milk powder	5–6	19	26	36

1.4. PRE-EMINENT CONSTITUENTS OF FOOD CHEMISTRY

Basic food chemistry generally deals with the 3 primary constituents in food: carbohydrates, proteins, and lipids.

1.4.1. Carbohydrates

Carbohydrates constitute a group of chemical compounds discovered in animal and plant cells. They possess an empirical formula $(CH_2O)_n$ or $C_nH_{2n}O_n$. As this formula is a combination of water and carbon (C) these materials are known as "carbohydrates or hydrates of C" (Weeks et al., 2004; Jakobek,

2015). Carbohydrates are the main product of the plant *photosynthesis* and are used up as fuel by animals and plants. Food carbohydrates comprise simple carbohydrates and complex carbohydrates.

1.4.2. Lipids

Lipids include oils, waxes, fats, and cholesterol. In the body, the fat assists as the main source of energy, a cushion nearby organs, and a thermal insulator and it is a significant constituent of the cell. As fats possess 2.25 times the energy content of the proteins and carbohydrates, most people struggle to restrict their consumption of dietary fat to avert becoming overweight. In most examples, *fats* are from animal products-milk products, eggs, seafood, and meats and *oils* are normally from plants-nuts, seeds, and olives. Lipids are used for flavor, to develop the quality of foods and to cook foods (Franzen and Kinsella, 1974; Becker, 2007).

1.4.3. Proteins

Proteins are vital components of food. Every cell needs protein for function and structure. Proteins are intricate polymers made up of amino acids. There are twenty amino acids discovered in the body. Eight out of twenty are vital for children and adults, and nine are crucial for infants. *Essential* means that one can't synthesize them in quite large quantities for repair and growth of the bodies, and thus, they should be included in the diet (Foster-Schubert et al., 2008; Pleissner et al., 2013). Proteins comprise long chains of around 100 to 500 amino acids that generally form into 3-D structures, which is their native state. When the native state of a protein is changed the 3-D structure also changes which are known as denaturation. Factors that trigger denaturation include acid, heating, beating, and freezing (Augustin et al., 2001).

1.5. PRE-EMINENT TERMS IN FOOD CHEMISTRY

1. **Amino acids** comprise C, nitrogen, hydrogen, and occasionally sulfur and assist as monomers to make proteins and peptides. Amino acids have the basic structure that comprises a carboxyl group (COOH) and an amino group (NH_2) attached to the C atom. There are twenty amino acids discovered in the body.
2. **Amylase** is an enzyme in saliva that is responsible for the breakdown of starch.

3. **An antioxidant** is a substance that slows down or prevents oxidation; prevents reactions encouraged by oxygen; frequently utilized as a preservative.
4. **Carbohydrate** is a compound of water and C. Carbohydrates are the most plentiful of all C comprising compounds, constituting almost three-fourths of the dry mass of the plant life on the earth. It is the main product of photosynthesis. Examples: glucose, starch, sucrose (table sugar) and cellulose.
5. **Casein is** a protein obtained from milk.
6. **Coagulation** is the transformation of liquid into a solid or soft mass.
7. **Denatured** is a change from the natural state. In the denatured protein, its properties or characteristics have been changed in some way, by chemicals, heat, or some enzymatic action, consequential in loss of the biological activity.
8. **Digestion** is the chemical breakdown of very large food compounds into relatively smaller molecules that can easily be absorbed by intestines in the animals and humans. The smaller molecules of food travel in the blood and are utilized by cells to produce the energy required by the body or to make other components.
9. **The emulsion** is a property where two liquids are spread out evenly in one another, however not dissolved in one another. Water and oil form the most communal emulsions and milk is generally an emulsion of the butterfat in water.
10. **Enzymatic browning** is a biochemical procedure in which vegetable or fruit tissues turn usually brown when exposed to oxygen. This procedure is catalyzed by the polyphenol oxidase.
11. **Enzymes** are protein catalysts, which govern particular chemical reactions in the living systems. Enzymes are normally active at low concentrations.
12. **Ester bonds** are bonds between glycerol molecules and fatty acids.
13. **Fatty acids** are building blocks of the fats, having atoms of hydrogen attached to the chains of C © atoms and oxygen comprising carbonyl group (-C=O). Fatty acids are discovered in almost every cell of the human body.

14. **Glycerol** is the backbone for the fatty acid molecules.
15. **Glucose** is a simple sugar and the main source of energy for several plants and mammals. It is also called dextrose, corn sugar, and grape sugar.
16. **Hormones** are substances that can impact enzyme action, physiology, and metabolism.
17. **Insoluble** means not capable to get dissolved. Fats are usually insoluble in water. Conversely, sugar is soluble in the water until more amounts are added than what a certain volume of water (H_2O) can dissolve.
18. **Lecithin** is an emulsifier found in soybean oil and eggs.
19. **Lipids** are compounds commonly of glycerol and fatty acids. Lipids are the most effective source of fuel in living things. Food lipids are generally divided into:

- **Fats**, which normally come from the sources of animals and are usually solid at room temperature.
- **Oils**, which normally come from the sources of plant and are usually liquid at room temperature.

20. **Melanin** is a group of black or brown pigments occurring in animals and plants.
21. **Organic** is associated with a branch of chemistry normally dealing with compounds of C. Though all of the living things comprise C and therefore are reflected to be organic, other compounds containing C have been yielded in the laboratory.
22. **Peptide bonds** are covalent bonds amongst two molecules of amino acid.
23. **Peptides** are short chains of amino acids.
24. **Photosynthesis** is a procedure by which the plant utilizes C dioxide and water to yield a simple sugar and oxygen. Plants store surplus sugar as starch.
25. **Polymers** usually contain 10 or more monomers. Starch is a polymer of monomer glucose. A protein is the polymer of amino acids.
26. **Polyphenol oxidase** is an enzyme-containing C, also known as phenolase, which catalyzes the oxidation of the phenolic compounds confined in plant tissue.

27. **Polyunsaturated** are fatty acids that possess multiple double bonds.
28. **Proteins** are complex polymers made up of amino acid monomers. Some instances of protein are hair, skin, muscle, hormones, and enzymes.
29. **Rennin** is an enzyme utilized to make cheese.
30. **Shortening** is a crystalline form of the solid fat.
31. **Soluble** means capable of getting dissolved. Solids or gases that dissolve are known as **solutes,** whereas the liquid involved in dissolving is known as a **solvent**. Like substances are normally soluble in the like solvents.
32. **Starch** is a polymer of glucose. It is an intricate carbohydrate discovered in green plants and a vital source of energy for humans and animals.
33. **Triacylglycerol** is a lipid compound comprising of 3 fatty acids connected to 1 glycerol molecule. This compound is a significant energy source for the human body.
34. **Vegetable oils** generally come from the plants like soybeans or corn and are a significant source of the polyunsaturated fats.

1.6. TACTIC TO STUDY OF THE FOOD CHEMISTRY

It is anticipated to develop an analytical tactic to the chemistry of the formulation, processing, and storage stability of the food so that details obtained from the study of one food or model system can improve the understanding of various other products (Accum, 1966; Acker, 1969). There are four components to this method:

(a) Determining the properties that are significant characteristics of high-quality and safe foods;

(b) Determining the biochemical and chemical reactions that have significant impacts on the loss of the quality or wholesomeness of the foods;

(c) Integrating the two points mentioned above so that an understanding is developed how the key biochemical and chemical reactions impact quality and safety; and

(d) Applying the understanding to several situations faced during food formulation, processing, and storage.

1.6.1. Safety and Quality Attributes

It is mandatory to repeat that safety is the first obligation of any food. In a broader sense, this means that the food should be free of any injurious chemical or microbial pollutant at the time of consumption. For operational objectives, this definition adopts a more practical form (Anonymous, 1831; Beaumont, 1833). In the canning industry, commercial sterility as applied to the foods having a low quantity of acid means the nonexistence of feasible spores of the *Clostridium botulinum.* This can generally be translated into a particular set of heating circumstances for a particular product in the specific package. Given the heating requirements, particular time-temperature conditions can be selected that will enhance the retention of the quality attributes (Clydesdale and Francis, 1977).

1.6.2. Biochemical and Chemical Reactions

Numerous reactions can modify food safety or quality. Every reaction class can comprise different substrates or reactants depending on the particular food and the specific circumstances for handling, processing, and storage (Eichner and Karel, 1972). They are dealt as reaction classes as the general nature of reactants or substrates is similar for all of the foods. Therefore, non-enzymic browning includes the reaction of the carbonyl compounds, which can ascend from prevailing decreasing sugars or from different reactions, like oxidation of the ascorbic acid, oxidation of lipids or hydrolysis of starch. Oxidation might involve lipids, vitamins, pigments, or proteins, and more explicitly, oxidation of lipids might include triacylglycerols in food or phospholipids in the other food (Davy, 1813, 1936).

1.6.3. Impact of Reactions on Safety and Quality of Food

Deterioration of food normally comprises of series of main events followed by the secondary events, which, become obvious as transformed quality attributes. Note mainly that the given attribute of quality can be varied as an outcome of different primary events (Filby, 1934; Erlandson and Wrolstad, 1972).

1.6.4. Analysis of Situations faced During Processing and Storage of Food

Having the description of attributes of safe, high-quality foods, the important chemical reactions included in deterioration of the food, and the association between the two, one can now initiate to ponder how to apply the information to situations faced during the processing and storage of food (Ihde, 1964; Hall, 1982). Temperature is possibly the most significant of these variables due to its broad impact on all kinds of chemical reactions. The impact of temperature on the reaction can be projected from the Arrhenius equation, k = Ae-D E/RT. Data imitating the Arrhenius equation produce a straight line when the log$k$ is plotted v. s 1/T. Arrhenius plots in Figure 1.2 signify reactions significant in food deterioration. It is obvious that food reactions usually conform to the Arrhenius relationship over the restricted intermediate temperature range but the deviations from the association can take place at low or high temperatures (McWeeny, 1968). Deviations from this Arrhenius relationship can take place due to the following events:

(a) Enzyme activity might be lost.
(b) The reaction pathway might change or might be impacted by competing reactions.
(c) One or more than one reactant might become depleted.
(d) The physical state of a system might change.

During storage of the food product, one often wants to find out how long the product can be anticipated to preserve a stated level of quality. Thus, one is more concerned about time with respect to integral of microbiological or chemical changes that take place during the specified storage period and in the method, these variations combine to find out the stated storage life for the food product. During the processing of food, one is frequently more concerned about the time it takes to deactivate a specific population of the microorganisms or how much time it takes for the reaction to progress to a quantified extent (LaJollo et al., 1971).

Another variable pH impacts the rates of several enzymic and chemical reactions. Extreme values of pH are usually needed for a severe reserve of enzymic processes or microbial growth and these conditions can outcome in an acceleration of base- or acid-catalyzed reactions. On the contrary, even a comparatively small change of pH can trigger profound variations in the quality of foods, for instance, muscle. The composition of a food product is significant as this defines the reactants accessible for chemical transformation (Stare and Whelan, 1978). Particularly important from the quality viewpoint is the association that prevails between the composition of raw material and the composition of the finished product. For instance,

(a) Manner in which vegetables and fruits are handled after harvesting can impact sugar content, and thus, impacts the degree of browning attained during deep-fat frying or dehydration.

(b) The manner in which the animal tissues are dealt with after dead cells impacts the rates and extents of ATP and glycolysis degradation and these can impact storage life, toughness, water-holding capacity, color, and flavor.

(c) The amalgamation of raw materials might cause unanticipated interactions; for instance, the oxidation rate can be inhibited or accelerated depending on the quantity of salt present.

Another significant compositional factor of reaction rates in the foods or food products is water activity (a_w). Many investigators have exhibited a_w to strongly impacts the rate of lipid oxidation, enzyme-catalyzed reactions, non-enzymic browning, chlorophyll degradation, sucrose hydrolysis, anthocyanin degradation, and several others. Most reactions incline to reduce in a rate below a_w corresponding to a range of the intermediate moisture foods (IMF). Oxidation of the lipids and related secondary effects, like carotenoid decoloration, are exemptions to this rule. These reactions speed up at the lower end of a_w scale (McCollum, 1959; Schoebel et al., 1969; Mayer, 1975).

More freshly, it has become obvious that glass transition temperature (Tg) of food and conforming water content (*W*g) of food at transition temperature are causatively linked to rates of the diffusion-restricted events in food. Therefore, Tg and *W*g have significance to physical properties of the dried and frozen foods, to conditions suitable for freeze-drying, to the physical changes including crystallization, gelatinization, recrystallization, and starch retrogradation. In the fabricated foods, the composition can be governed by adding accepted chemicals, like acidulants, chelating agents, antioxidants or flavors, or by eliminating unwanted reactants, for instance, eliminating glucose from the dehydrated egg albumen (Taylor, 1980; Whelan, 1993).

The composition of the atmosphere is significant primarily to comparative oxygen content and humidity, even though CO_2 and ethylene are also vital during the storage of the living plant foods. Unfortunately, in circumstances where elimination of oxygen is anticipated, this is almost difficult to achieve completely. The detrimental concerns of a small amount of residual oxygen occasionally become obvious during product storage. For instance, the early formation of the small amount of dehydroascorbic acid can lead to the

Maillard browning during the stage of storage. For some of the products, exposure to the light can be harmful, and it is then suitable to package the food products in light-resistant material or to govern the wavelengths and intensity of light, if feasible. Food chemists should be able to incorporate information regarding quality attributes of the foods, deteriorative reactions to which the foods are vulnerable and the factors controlling rates and kinds of these reactions, to solve issues linked to the formulation, processing, and storage of food (Schoebel et al., 1969; Quast and Karel, 1972).

1.7. SOCIETAL ROLE OF THE FOOD CHEMISTS

1.7.1. Why Should these Chemists Become Indulged in Societal Issues?

For the reasons given below, food chemists should feel appreciative to become indulged in societal problems that incorporate pertinent technological aspects.

1. Food chemists had the opportunity of receiving high-level education and obtaining special scientific skills and these skills and privileges carry with them a conforming high level of obligation.
2. Activities of the food chemist's impact competence of the food supply, cost of foods, the healthfulness of population, waste disposal and creation, water, and energy utilization and nature of the food regulations. As these matters intrude on the welfare of the public, it is realistic that the food chemists must feel an obligation to have the activities focused on the advantage of society.

If food chemists don't become indulged in the techno-societal issues, the thoughts of others— professional lobbyists, scientists from the other professions, persons in news media, charlatans, and consumer activists— will prevail.

1.7.2. Types of Involvement

The societal responsibilities of food chemists comprise good job performance, protecting the morals of the scientific community, good citizenship, but the gratification of these necessary roles isn't enough. An extra role of great significance, and one that frequently goes unfulfilled by the food chemists, is of helping decide how scientific knowledge is understood and utilized by society.Even though food chemists and the other food scientists must not have the input to these conclusions, they should, in the importance of sensible decision making, have their opinions heard and considered. Acceptance of this post, which is certainly unquestionable, leads to the apparent question, "What accurately must food chemists do to correctly discharge their duties in this regard?" Several activities are suitable.

1. Participate in pertinent professional societies;
2. When invited serve on the governmental advisory committee; and
3. Undertake personal advantages of public service nature.

In summary, scientists generally have greater responsibilities to society than do individuals without scientific education. Scientists are anticipated to produce knowledge in an ethical and productive manner, but this isn't enough. They must also accept the obligation of confirming that scientific knowledge is utilized in a way that will produce the greatest advantage to society. Fulfillment of this responsibility needs that scientists must not only struggle for brilliance and conformance to the high ethical standards in the professional activities, but they must also cultivate a deep-seated worry for the scientific enlightenment and well-being of the public.

REFERENCES

1. Accum, F., (1966). *A Treatise on Adulteration of Food, and Culinary Poisons, 1920* (Vol. 1, pp. 1–40), Facsimile reprint by Mallinckrodt Chemical Works, St. Louis, MO.
2. Acker, L. W., (1969). Water activity and enzyme activity. *Food Technol., 23*(10), 1257–1270.
3. Aletor, O. L. U. W. A. T. O. Y. I. N., Oshodi, A. A., and Ipinmoroti, K., (2002). Chemical composition of common leafy vegetables and functional properties of their leaf protein concentrates. *Food Chemistry, 78*(1), 63–68.
4. Amarteifio, J. O., and Moholo, D., (1998). The chemical composition of four legumes consumed in Botswana. *Journal of Food Composition and Analysis, 11*(4), 329–332.
5. Andrée, S., Jira, W., Schwind, K. H., Wagner, H., and Schwägele, F., (2010). Chemical safety of meat and meat products. *Meat Science, 86*(1), 38–48.
6. Andrés-Lacueva, C., Medina-Remon, A., Llorach, R., Urpi-Sarda, M., Khan, N., Chiva-Blanch, G., and Lamuela-Raventos, R. M., (2010). Phenolic compounds: Chemistry and occurrence in fruits and vegetables. *Fruit and Vegetable Phytochemicals: Chemistry, Nutritional Value, and Stability, 1*, 53–80.
7. Anonymous, (1831). Death in the pot. Cited by Filby, 1934 (Vol. 1, pp. 1–30).
8. Augustin, M. A., Sanguansri, L., Margetts, C., and Young, B., (2001). Microencapsulating food ingredients. *Food Australia, 53*(6), 220–223.
9. Beaumont, W., (1833). *Experiments and Observations of the Gastric Juice and the Physiology of Digestion* (Vol. 1, pp. 1–34). F. P. Allen, Plattsburgh, NY.
10. Becker, E. W., (2007). Micro-algae as a source of protein. *Biotechnology Advances, 25*(2), 207–210.
11. Belitz, H., Grosch, W., and Schieberle, P., (2004). *Food Chemistry* (Vol. 1, pp. 1–49). Springer Verlag, Berlin-Heidelberg.
12. Benjakul, S., Visessanguan, W., Phongkanpai, V., and Tanaka, M., (2005). Antioxidative activity of caramelization products and their preventive effect on lipid oxidation in fish mince. *Food Chemistry, 90*(1–2), 231–239.

13. Browne, C. A., (1944). *A Source Book of Agricultural Chemistry* (Vol. 1, pp. 1–35), Chronica Botanica Co., Waltham, MA.
14. Calvert, J. J., (1985). Food selection by western gorillas (Gg gorilla) in relation to food chemistry. *Oecologia, 65*(2), 236–246.
15. Chevreul, M. E., (1824). Considérations Générales Sur L'analyse Organique Et Sur Ses Applications (Vol. 1, pp. 1–29). Cited by Filby.
16. Choct, M., and Hughes, R. J., (1999). Chemical and physical characteristics of grains related to variability in energy and amino acid availability in poultry. *Australian Journal of Agricultural Research, 50*(5), 689–702.
17. Choct, M., (1997). Feed non-starch polysaccharides: Chemical structures and nutritional significance. *Feed Milling International, 191*(June issue), 13–26.
18. Clydesdale, F. M., and Francis, F. J., (1977). *Food, Nutrition, and You* (Vol. 1, pp. 1–51), Prentice-Hall, Englewood Cliffs, NJ.
19. Creamer, L. K., and MacGibbon, A. K., (1996). Some recent advances in the basic chemistry of milk proteins and lipids. *International Dairy Journal, 6*(6), 539–568.
20. Cui, W., (2005). *Food Carbohydrates* (Vol. 1, pp. 1–40). CRC Press, Boca Raton.
21. Davy, H., (1813). *Elements of Agricultural Chemistry, in a Course of Lectures for the Board of Agriculture* (Vol. 1, pp. 1–44), Longman, Hurst, Rees, Orme, and Brown, London, Cited by Browne, 1944.
22. Davy, H., (1936). *Elements of Agricultural Chemistry* (Vol. 5, pp. 1–50), Longman, Rees, Orme, Brown, Green, and Longman, London.
23. De Almeida, C. G. E., da Silva Queiroz-Monici, K., Reis, S. M. P. M., and de Oliveira, A. C., (2006). Chemical composition, dietary fiber and resistant starch contents of a raw and cooked pea, common bean, chickpea, and lentil legumes. *Food Chemistry, 94*(3), 327–330.
24. Demirezen, D., and Uruç, K., (2006). Comparative study of trace elements in certain fish, meat, and meat products. *Meat Science, 74*(2), 255–260.
25. Dombrowski, W. M., and Sikorski, Z., (2005). *Toxins in Food* (Vol. 1, pp. 1–41). CR Press, Boca Raton.
26. Eichner, K., and Karel, M., (1972). The influence of water content and water activity on the sugar-amino browning reaction in model systems under various conditions. *J. Agric. Food Chem. 20*(2), 218–223.

27. Enujiugha, V. N., (2003). Chemical and functional characteristics of conophor nut. *Pakistan Journal of Nutrition*, *2*(6), 335–338.
28. Erlandson, J. A., and Wrolstad, R. E., (1972). Degradation of anthocyanins at limited water concentration. *J. Food Sci., 37*(4), 592–595.
29. Fardet, A., Rock, E., and Rémésy, C., (2008). Is the in vitro antioxidant potential of whole-grain cereals and cereal products well reflected in vivo? *Journal of Cereal Science*, *48*(2), 258–276.
30. Fennema, O. R., (1985). *Food Chemistry* (Vol. 1, pp. 1–24.). Marcel Dekker Inc., New York-Basel.
31. Filby, F. A., (1934). *A History of Food Adulteration and Analysis* (Vol. 1, pp. 1–20), George Allen and Unwin, London.
32. Foster-Schubert, K. E., Overduin, J., Prudom, C. E., Liu, J., Callahan, H. S., Gaylinn, B. D., and Cummings, D. E., (2008). Acyl and total ghrelin are suppressed strongly by ingested proteins, weakly by lipids, and biphasically by carbohydrates. *The Journal of Clinical Endocrinology and Metabolism*, *93*(5), 1971–1979.
33. Franzen, K. L., and Kinsella, J. E., (1974). Parameters affecting the binding of volatile flavor compounds in model food systems. I. Proteins. *Journal of Agricultural and Food Chemistry*, *22*(4), 675–678.
34. Gray, J. I., Pearson, A. M., and Monahan, F. J., (1994). Flavor and aroma problems and their measurement in meat, poultry, and fish products. In *Quality Attributes and their Measurement in Meat, Poultry, and Fish Products* (Vol. 1, pp. 250–288). Springer, Boston, MA.
35. Gross, R., Koch, F., Malaga, I., De Miranda, A. F., Schoeneberger, H., and Trugo, L. C., (1989). Chemical composition and protein quality of some local Andean food sources. *Food Chemistry*, *34*(1), 25–34.
36. Hall, R. L., (1982). Food additives. In: Kirk, D., and Eliason, I. K., (eds.), *Food*, and *People* (Vol. 1, pp. 148–156). Boyd and Fraser, San Francisco.
37. Hudson, B. J. F., (1981). *Developments in Food Proteins* (Vol. 1–7, pp. 1–60), Elsevier Appl. Sci. Publ. London-New York.
38. Huppertz, T., and Kelly, A. L., (2006). Physical chemistry of milk fat globules. In *Advanced Dairy Chemistry Volume 2 Lipids* (Vol. 2, pp. 173–212). Springer, Boston, MA.
39. Ihde, A. J., (1964). *The Development of Modern Chemistry* (Vol. 1, pp. 1–34), Harper, and Row, New York.

40. Iwai, K., Hasegawa, T., Taguchi, Y., Morimatsu, F., Sato, K., Nakamura, Y., and Ohtsuki, K., (2005). Identification of food-derived collagen peptides in human blood after oral ingestion of gelatin hydrolysates. *Journal of Agricultural and Food Chemistry*, *53*(16), 6531–6536.
41. Jack, E. L., and Smith, L. M., (1956). Chemistry of milk fat: A review. *Journal of Dairy Science*, *39*(1), 1–25.
42. Jakobek, L., (2015). Interactions of polyphenols with carbohydrates, lipids, and proteins. *Food Chemistry*, *175*, 556–567.
43. Jukes, T. H., (1978). How safe is our food supply? *Arch. Intern. Med. 138*(1), 772–774.
44. Labuza, T. P., Tannenbaum, S. R., and Karel, M., (1970). Water content and stability of low-moisture and intermediate moisture foods. *Food Techol., 24*(5), 543–550.
45. LaJollo, F., Tannenbaum, S. R., and Labuza, T. P., (1971). Reaction at limited water concentration. Chlorophyll degradation. *J. Food Sci., 36*(6), 850–853.
46. Liebig, J. V., (1847). *Researches on the Chemistry of Food* (Vol. 1, pp. 1–34). Edited from the author's manuscript by William Gregory; London, Taylor, and Walton London. Cited by Browne, 1944.
47. Maga, J. A., (1982). Phytate: Its chemistry, occurrence, food interactions, nutritional significance, and methods of analysis. *Journal of Agricultural and Food Chemistry*, *30*(1), 1–9.
48. Mayer, J., (1975). A Diet for Living (Vol. 1, pp. 1–24). David McKay, Inc., New York.
49. McCollum, E. V., (1959). The history of nutrition. *World Rev. Nutr. Diet. 1*, 1–27.
50. McWeeny, D. J., (1968). Reactions in food systems: Negative temperature coefficients and other abnormal temperature effects. *J. Food Technol., 3*, 15–30.
51. Morton, I. D.,and MacLeod, A. J., (1986). *Food Flavors* (Vol. 1, pp. 1–29). Elsevier, Amsterdam-Oxford-New York-Tokyo.
52. Mosse, J., (1990). Nitrogen-to-protein conversion factor for ten cereals and six legumes or oilseeds. A reappraisal of its definition and determination. Variation according to species and to seed protein content. *Journal of Agricultural and Food Chemistry*, *38*(1), 18–24.
53. Mottram, D. S., (1998). Flavor formation in meat and meat products: A review. *Food Chemistry*, *62*(4), 415–424.

54. Mottram, D. S., Wedzicha, B. L., and Dodson, A. T., (2002). Food chemistry: Acrylamide is formed in the Maillard reaction. *Nature, 419*(6906), 448.

55. Ortega-Calvo, J. J., Mazuelos, C., Hermosin, B., and Sáiz-Jiménez, C., (1993). Chemical composition of spirulina and eukaryotic algae food products marketed in Spain. *Journal of Applied Phycology*, *5*(4), 425–435.

56. Otles, S., (2005). *Pesticide, Veterinary, and Other Residues in Food* (Vol. 1, pp. 1–27.). CRC Press, Boca Raton.

57. Palmer, L. S., (1924). The chemistry of milk and dairy products viewed from a colloidal standpoint. *Industrial and Engineering Chemistry, 16*(6), 631–635.

58. Pearson, A. M., (2013). *Quality Attributes and Their Measurement in Meat, Poultry, and Fish Products* (Vol. 9, pp. 1–20). Springer.

59. Piironen, V., Toivo, J., and Lampi, A. M., (2002). New data for cholesterol contents in meat, fish, milk, eggs, and their products consumed in Finland. *Journal of Food Composition and Analysis, 15*(6), 705–713.

60. Pleissner, D., Lam, W. C., Sun, Z., and Lin, C. S. K., (2013). Food waste as nutrient source in heterotrophic microalgae cultivation. *Bioresource Technology*, *137*, 139–146.

61. Quast, D. G., and Karel, M., (1972). Effects of environmental factors on the oxidation of potato chips. *J. Food Sci., 37*(4), 584–588.

62. Rasper, V. E., and Walker, C. E., (2000). Quality evaluation of cereals and cereal products. In *Handbook of Cereal Science and Technology, Revised, and Expanded* (Vol. 1, pp. 521–530). CRC Press.

63. Renaud, S. M., and Parry, D. L., (1994). Microalgae for use in tropical aquaculture II: Effect of salinity on growth, gross chemical composition and fatty acid composition of three species of marine microalgae. *Journal of Applied Phycology*, *6*(3), 347–356.

64. Renaud, S. M., Thinh, L. V., and Parry, D. L., (1999). The gross chemical composition and fatty acid composition of 18 species of tropical Australian microalgae for possible use in mariculture. *Aquaculture, 170*(2), 147–159.

65. Renaud, S. M., Thinh, L. V., Lambrinidis, G., and Parry, D. L., (2002). Effect of temperature on growth, chemical composition and fatty acid composition of tropical Australian microalgae grown in batch cultures.

Aquaculture, *211*(1–4), 195–214.

66. Risch, S. J., (2009). Food packaging history and innovations. *Journal of Agricultural and Food Chemistry*, *57*(18), 8089–8092.
67. Ross, A. B., Shepherd, M. J., Schüpphaus, M., Sinclair, V., Alfaro, B., Kamal-Eldin, A., and Åman, P., (2003). Alkylresorcinols in cereals and cereal products. *Journal of Agricultural and Food Chemistry*, *51*(14), 4111–4118.
68. Saini, R. K., Nile, S. H., and Park, S. W., (2015). Carotenoids from fruits and vegetables: Chemistry, analysis, occurrence, bioavailability, and biological activities. *Food Research International*, *76*, 735–750.
69. Schmidt, O., Quilter, J. M., Bahar, B., Moloney, A. P., Scrimgeour, C. M., Begley, I. S., and Monahan, F. J., (2005). Inferring the origin and dietary history of beef from C, N, and S stable isotope ratio analysis. *Food Chemistry*, *91*(3), 545–549.
70. Schoebel, T., Tannenbaum, S. R., and Labuza, T. P., (1969). Reaction at limited water concentration. (1) Sucrose hydrolysis. *J. Food Sci.*, *34*(4), 324–329.
71. Shukla, P., Sharma, A., and Sharma, A., (2017). Food additives from an organic chemistry perspective. *MOJ Bioorganic and Organic Chemistry. Review Article*, *1*(3), 70–79.
72. Sikorski, Z. E., and Kolakowska, A., (2004). *Chemical and Functional Properties of Food Lipids* (Vol. 1, pp. 10–30). CRC Press, Boca Raton.
73. Stadler, R. H., Blank, I., Varga, N., Robert, F., Hau, J., Guy, P. A. and Riediker, S., (2002). Food chemistry: Acrylamide from Maillard reaction products. *Nature*, *419*(6906), 449.
74. Stare, F. J., and Whelan, E. M., (1978). *Eat OK—Feel OK,* Christopher Publishing House, North Quincy, MA. *1*, 1-54.
75. Sundram, K., Sambanthamurthi, R., and Tan, Y. A., (2003). Palm fruit chemistry and nutrition. *Asia Pacific Journal of Clinical Nutrition*, *12*(3), 1–20.
76. Taylor, R. J., (1980). *Food Additives,* John Wiley and Sons, New York. *1*, 1-19.
77. Tsao, R., (2010). Chemistry and biochemistry of dietary polyphenols. *Nutrients*, *2*(12), 1231–1246.
78. Van, B. J. P., (1979). The chemistry of texture in fruits and vegetables. *Journal of Texture Studies*, *10*(1), 1–23.

79. Vrabcheva, T., Usleber, E., Dietrich, R., and Märtlbauer, E., (2000). Co-occurrence of ochratoxin A and citrinin in cereals from Bulgarian villages with a history of Balkan endemic nephropathy. *Journal of Agricultural and Food Chemistry*, *48*(6), 2483–2488.
80. Weeks, R. D., Wilson, L. T., Vinson, S. B., and James, W. D., (2004). Flow of carbohydrates, lipids, and protein among colonies of polygyne red imported fire ants, Solenopsis invicta (Hymenoptera: Formicidae). *Annals of the Entomological Society of America*, *97*(1), 105–110.
81. Weidenborner, M., (2001). *Encyclopedia of Food Mycotoxins,* Springer Verlag, Berlin-Heidelberg. *1*, 10–43.
82. Whelan, E. M., (1993). *Toxic Terror* (Vol. 1, pp. 1–28), Prometheus Books, Buffalo, NY.
83. Wolff, J., (2000). Ochratoxin A in cereals and cereal products. *Archives for food hygiene*, 51(4/5), 85–88.
84. Zeleny, L., and Coleman, D. A., (1938). Acidity in cereals and cereal products its determination and significance. *Cereal Chemistry*, 15, 580–595.

Chapter 2

Characteristics of Major Food Constituents

CONTENTS

2.1. INTRODUCTION

Raw food products obtained from animals and plants contain mostly water, protein, carbohydrate, and lipid as major components. Understanding the fundamental characteristics such as the physicochemical properties and composition of the raw food products during the stage of handling, processing, and storage stability will be very significant (Biliaderis, 1991; Gülçin, 2012).

Water is a crucial constituent in which the life processes of organisms occur. In food, water signifies mainly as an extracellular or intracellular component in animal and plant food products. Contact of water with some other food constituent's outcomes in controlling many physical and chemical reactions which considerably contributes to quality and shelf life values of the food products (Kwon et al., 1965; Luck, 1981). For example, elimination of water from the food products by chemical or physical methods like binding it with sugar or salt inhibits enzymatic and chemical reactions produced by the microorganisms, consequential in shelf life enhancement of the food products (Lewin, 1974; Luck, 1981). Conversely, physical contact of water with the macromolecules might contribute to the physical appearance of the food.

Table 2.1: Properties of Water (H_2O) at Diverse Temperature Conditions

Water	Temperature (C)					
	0	20	40	60	80	100
Vapor pressure (mm Hg)	4.58	17.53	55.32	149.4	355.2	760.0
Specific heat (cal/g C)	1.0074	0.9988	0.9980	0.9994	1.0023	1.0070
Density (g/cm^3)	0.9998	0.9982	0.9922	0.9832	0.9718	0.9583
Thermal conductivity (kcal/m^2 h C)	0.486	0.515	0.540	0.561	0.576	0.585
Heat of vaporization (cal/g)	597.2	586.0	574.7	563.3	551.3	538.9
Viscosity (centipoises)	1.792	1.002	0.653	0.466	0.355	0.282
Surface tension (dynes/cm)	75.62	72.75	55.32	69.55	66.17	62.60
Dielectric constant	88.0	80.4	73.3	66.7	60.8	55.3
Refractive index	1.3338	1.3330	1.3306	1.3272	1.3230	1.3180
Coefficient of thermal expansion x 10^4	–	2.07	3.87	5.38	6.57	–

Food lipids are usually non-soluble compounds referred to as oils (liquid) or fats (solid) demonstrating their physical state at the ambient temperatures. Food lipids are classified as polar and nonpolar lipids to specify differences in their functional properties and solubility. Food lipids generally play a significant role in the food sensory quality by backing to attributes like texture, flavor, caloric density, and nutrition (Pérez-Jiménez and Saura-Calixto, 2006).

Proteins play a vital role in food and biological systems as well. All proteins derived biologically can be utilized as food proteins which should be easily digestible, nutritionally adequate, nontoxic, functionally usable in the food products, sustainable agriculturally, and available in abundance. The useful properties of proteins are linked to their physicochemical and structural characteristics (Korycka-Dahl et al., 1978; Aasen et al., 2003).

Carbohydrates are the organic compounds that are normally most abundant and distributed on earth. They consist of hydrogen, oxygen, and carbon (C) and can be complex or simple in the structure of the molecules. In the food processing, carbohydrate plays a useful role in its chemical and physical properties which usually act as sweeteners, stabilizers, gelling agents, thickeners, precursors, and fat replacers for coloring substances and aroma, particularly in thermal processing. Carbohydrates are normally divided into monosaccharides, polysaccharides, and oligosaccharides (Lieberman et al., 1982; Mazzio et al., 1998).

Enzymes are proteins included in chemical reactions linked to metabolism. Food enzymes can normally be classified as exogenous and endogenous. Endogenous enzymes are the enzymes that are existent and synthesized by the biological cells in all of the organisms and therefore deteriorate or enhance food quality, whereas exogenous enzymes are the enzymes that are usually added to the foods to trigger a desirable change (Damodaran et al., 2008).

Inactivation or governing the rate of the enzyme is important to outcome in anticipated changes in the food products (Escarpa et al., 1997; Barsotti et al., 2001).

In this chapter, properties of water, protein, carbohydrate, enzyme, and lipid such as their chemical and physical properties, chemical structure, and useful properties in the food are discussed.

2.2. WATER

2.2.1. Physical Properties of Ice and Water

Physical properties of the water are displayed in Table 1; considerable information is obtained from De Man (1999). Table 2.1 displays the physical properties of the ice at several temperatures. Some physical properties of ice and water are exceptional; though, these extraordinarily high values of caloric properties of the water would be significant for the food processing operations like freezing and drying (Fennema, 1973; Mala and Li, 1999). In frozen food, the density difference between ice and water might contribute to the structural damage to foods. Structural change from solids to semisolids or vice versa as an outcome of changing temperature will also change the density of ice, yielding structural damage and stresses in the frozen foods (Boutron et al., 1960; Vaclavik and Christian, 2014).

Table 2.2: Physical Properties of Ice

Ice	**Temperature (C)**						
	0	**5**	**10**	**15**	**20**	**25**	**30**
Vapor pressure (mm Hg)	4.58	3.01	1.95	1.24	0.77	0.47	0.28
Heat of sublimation (cal/g)	677.8	–	672.3	–	666.7	–	662.3
Heat of fusion (cal/g)	79.8	–	–	–	–	–	–
Specific heat (cal/g C)	0.4873	–	0.4770	–	0.4647	–	0.4504
Density (g/cm^3)	0.9168	0.9171	0.9175	0.9178	0.9182	0.9185	0.9188
Heat capacity (joule/g)	2.06	–	–	–	1.94	–	–
Coefficient of thermal expansion x 10^5	9.2	7.1	5.5	4.4	3.9	3.6	3.5

2.2.2. Interactions of Water

2.2.2.1. *Water-Solute Interaction*

Interaction of Water-solute results in the modification of properties of food systems. These changes are caused by the molecular interactions amongst the nature of the solute and the water, mainly from charged groups or ions which intermingle with water through the electrostatic forces. These interactions change the geometric structure of the water molecules available in hydrophilic groups of the solutes which outcome in their reactivity and mobility retardation (Brady and Romanov, 1960). On the other hand, hydrophobic groups of the solutes weakly interact with nearby water, because of their inclination to the non-aqueous surrounding environment (Peng et al., 1994a, b; Ashbaugh et al., 2002). Though, this weak interaction might result in structural consequences and instability.

2.2.2.2. *Types of Water*

Most of the natural fresh foods comprise water up to 70% of their weight; vegetables and fruits comprise water up to 95% or greater (Vaclavik and Christian, 2014). Water presents in the food in the form of bound, free, and entrapped water (Fennema, 1973; Schmidt, 2004). Free water is the water that can be eliminated easily from foods by squeezing, pressing, or grinding, for example, removing water from the orange fruit by squeezing it.

Bound water is the water that can't be easily extracted since it binds tightly onto the water molecules (Fennema, 1973). Therefore, this kind of water requires particular analysis approaches to determine the value in a particular food.

Water might also be present as entangled water in foods like water in the pectin gel or vegetable or fruit. Entrapped water is halted in matrixes, cells, or capillaries but if unrestricted during damage or cutting, it flows freely. Entrapped water possesses properties of free water. Entrapped water in the food is often linked with the water holding capacity (WHC), a term frequently employed to define the ability of matrix of the molecules, normally macromolecules existent at low concentration, to entrap physically large quantities of water in a way that prevents exudation under the use of an external force, often gravitational force (Belitz et al., 2009; Damodaran et al., 2008; De Man, 1999). Acquainted food matrices that capture water in this method include gels of starch and pectin and cells of the tissues, both animal and plant (Mala et al., 1997; Hwang et al., 2009).

Physically entrapped water doesn't flow readily from the tissue foods, even when these tissue foods are minced or cut. However, this water performs during the processing with properties near to those of the pure water. It is easily eliminated during drying, easily transformed into ice during the freezing process and readily accessible as a solvent. Therefore, yet its bulk flow is strictly controlled, the movement of distinct molecules is fundamentally similar to that of the water molecules in the dilute salt solution (Fennema, 1973, 1996).

Most water in gels and tissues can be classified as physically entrapped and damage to this entrapment capability of foods has a profound effect on food quality. Examples of the quality defects linked with damage to the WHC are thaw exudation from the previously frozen foods, syneresis of gels, and poorer performance of the animal tissue in sausage consequential from the deterioration in pH of muscle that complements normal physiological postmortem events. In all circumstances, the quality defect stanches from the physical rearrangement of the molecules of water in space but doesn't certainly reflect any substantial change in the interactive properties of the molecules (De Man, 1999; Schmidt, 2004).

2.2.2.3. Water as Dispersing Medium and Food Solvent

Solutes in the form of the food components can be dispersed, suspended, or dissolved in water depending on the particle solubility and size. To make a solution, water must dissolve small molecules like salts, water-soluble vitamins, or sugars which might be either molecular or ionic. An ionic solution is made by dissolving the substances that generally ionize in water, like salts, bases, or acids. Na+ chloride comprises chloride (Cl–) and sodium (Na+) ions detained together by the ionic bonds; the additional water molecules decrease the attractive forces amongst oppositely charged ions, then the ionic bonds are released, and therefore the water molecules cover the distinct hydrated ions. Each ion is normally surrounded by 6 water molecules and the ions move freely of each other (Law et al., 1980; Kim et al., 2006; Hyung and Kim, 2008).

The surrounding idea of water molecules also takes place in the polar molecules to form molecular solutions, for example, sugars. When the sugar crystal is normally dissolved in water, major hydrogen bonds amongst polar hydroxyl groups are relaxed and substituted by the hydrogen bonds amongst the sugar molecules and water. Thus, the hydration takes place slowly until every molecule of sugar is surrounded by water molecules. Temperature also disturbs the hydrogen bond disturbance amongst polar hydroxyl groups of the sucrose molecules, then assisting the creation of hydrogen bonds amongst sucrose and water and hydration of the sucrose molecules. Thus, sucrose has better solubility in the hot water than that in cold water. Solute dissolution in the water will upsurge boiling point of the water which is important to be understood for the food process application such as in jelly or candy making through the determination of sucrose absorption by simply measuring boiling point of the sucrose solution (Fennema, 1996; Al-Muhtaseb et al., 2002).

Big molecules having the size range 1 to 100 nm are hard to dissolve fully to make true solutions, however, they can be dispersed to create colloidal dispersion. Examples of such kind of molecules include starches, hydrocolloids, and some of the food proteins. Colloidal dispersions that generally apply in foods are frequently unstable systems; stabilization is essential to sustain water-solute binding, therefore phase separation can normally be prevented. They are mainly unstable to factors like freezing, heating, or pH change. In food applications, varying the circumstances in the stable dispersion can trigger the system to become unstable like precipitation or gelation; nevertheless, this is needed in some cases, for example, when making the pectin jellies (Belitz et al., 2009). Colloidal dispersion is crucial to be understood in the food processing as numerous packaged or convenient foods possess colloidal dimensions and their sensitivity and stability to certain kinds of reactions (Cadden, 1988; Velázquez and Feirtag, 1999).

Particles that are quite large, normally >100 nm, will create a suspension when assorted with water. The suspension is more unstable, which splits or divides out over a period as compared to colloidal dispersion. An instance of suspension is when the native grains of starch is suspended in water (Wang and Brennan, 1991; Al-Muhtaseb et al., 2004).

Table 2.3: The Water Activity of Certain Foods

Food	Water activity (a_w)
Leberwurst	0.96
Dried fruits	0.72–0.80
Salami	0.82–0.85

Honey	0.75
Marmalades	0.82–0.94

2.2.3. Water and Food Stability

As water is a main compound in the food, it has an important role to govern food stability. The stability of food comprises its chemical, microbiological, and physical aspects linked with food quality; these parameters will relate to and affect each other. Food instability can trigger food perishability which damages food quality. It has been recognized that there exists an association between food perishability and water content (Boutron et al., 1986; Duckworth et al., 1976). In some of the foods, water is an unwanted compound. Dehydration and concentration processes are performed mainly for the aim of reducing the water content of food, at the same time increasing the concentration of the solutes and thus decreasing perishability (Scott, 1957; Patino et al., 2008).

Interestingly, it has been investigated that several kinds of food with similar water content have diverse significant perishability. Thus, it is obvious that the sign of perishability isn't only the water content (Soottitantawat et al., 2004; Xiao et al., 2012). As water can intermingle with solutes, the intensity binding may expect that water involved in strong relations to trigger food quality degradation primarily by hydrolytic chemical reactions and microorganism growth. The term water activity was presented to reveal the intensity with which the water links with several solute constituents (Luck, 1981; Scott, 1957).

Even though water activity isn't a completely reliable factor in food safety and stability, this can give enough information about its correlation with the rates of microbial growth and with the rates of numerous degradative reactions. Water activity is reflected high in the living tissues of organisms as they need an adequate level of water to sustain cell metabolism. However, microorganisms like bacteria, yeast, and mold multiply at a high water activity (Labuza et al., 1972; Kaymak-Ertekin and Sultanoğlu, 2001). Preservation methods against spoilage due to these microorganisms consider governing of the water activity in food. Less bacterial growth takes place if the level of water is dropped to less than 0.85 conferrings to FDA Model Food Code. Foods with water activity values among 0.6 and 0.9 are called intermediate moisture foods (IMF) (Vaclavik and Christian, 2014).

Water activity isn't the only factor that backs to bacterial growth; other factors like environment and pH would also contribute to food safety and stability. These foods are essential to be secured against microbial spoilage. In the food application, a choice for reducing water activity and therefore enhancing the food's shelf life is to utilize additives having high water binding capacities, like sucrose, salt, glycerol, and sorbitol, which work as humectants (Fennema, 1978). For examples, jams, dried fruits, and jellies are prepared to utilize high concentrations of brines and sugar, while hams, sausages, and salami comprise high concentrations of salt to increase their shelf life (Angibaud, 1986; Ferry, 1980; White and Cakebread, 1966; White and Cakebread, 1969; Williams et al., 1955).

The storage food quality is dependent on water activity and not on the content of water. The ratio of moisture content of the food and the comparative humidity of surrounding air is called the water activity. It is a vital characteristic. Water activity is determined by (Vaclavik and Christian, 2014).

$$Aw = P/P_0 = ERH \quad (1)$$

Where, $P \rightarrow$ partial vapor pressure of the food moisture at temperature T_0; $P_0 \rightarrow$ saturation vapor pressure of the pure water (H_2O) at temperature T_0; $T_0 \rightarrow$ equilibrium temperature of the system; ERH $\rightarrow$equilibrium relative humidity.

An association between ERH and Aw is noteworthy. First, water activity is a fundamental property of the sample, while percentage ERH is the property of atmosphere developed in the steady-state with a sample. Second, equality prevails in Eq. (1) only if the equilibrium has been reached or developed between the product and environment. Equilibrium condition of samples is quite a time-consuming process and nearly difficult for large samples, particularly at temperatures below 20°C (Damodaran et al., 2008; Shao and Li, 2013).

2.3. LIPIDS

2.3.1. Chemical Structure and its Classification

Oils and fats are triglycerides, the main constituent of lipids. Overall, a lipid is a term that comprises the triglycerides, sterols, and phospholipids (Slade and Levine, 1995). Glycerides comprise monoglycerides, diglycerides, and triglycerides. Glycerides have the glycerol molecule backbone linked to one or more than one fatty acid molecules (O'Keefe, 2002). A monoglyceride comprises glycerol esterified to usually one fatty acid molecule. Diglycerides are created by esterification of 2 fatty acids to glycerol. If the triglyceride comprises 3 same fatty acids, it is known as a simple triglyceride, else if it comprises 2 or 3 diverse fatty acids it is known as the mixed triglyceride (O'Keefe, 2002; Weiss, 1983). The arrangement and particular kind of fatty acids on glycerol conclude the physical and chemical properties of a fat (Serdarevich, 1967; Zilversmit, 1968).

In addition to free fatty acids and glycerides, lipids might comprise small amounts of tocopherols, phospholipids, sterols, vitamins, and fat-soluble pigments (Nielsen, 2003). Phospholipids are analogous to triglycerides but comprise of only 2 fatty acids esterified to glycerol. A polar group comprising phosphoric acid and the nitrogen-containing group replaces the third fatty acid. Sterols comprise a steroid nucleus, an 8–10 C side chain and an alcohol group (Robinson and Eglinton, 1990; Giese, 1996).

Fatty acids are generally long hydrocarbon chains having CH_3 (methyl) group at one end and the COOH (carboxylic acid) group at the other end of the chain. Most of the natural fatty acids comprise from four to 24 C atoms and normally an even number of C atoms in the chain; for example, butyric acid is usually the smallest fatty acid, comprising four C atoms (Murakami et al., 1996; Seu et al., 2006).

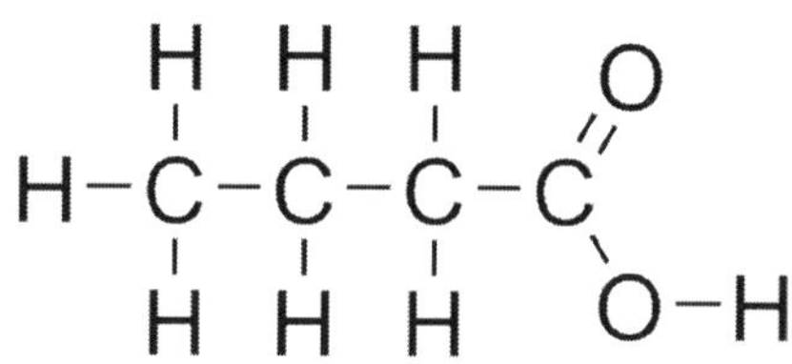

Figure 2.1: Butyric acid- an instance of saturated fatty acid.

Source: https://allnutriments.blogspot.com/2013/04/saturated-fatty-acid-structure-definition.html.

Fatty acids might be saturated, in the case, they comprise only single C-to-C bonds and possesses the general formula CH_3 $(CH_2)_n$COOH (Fennema, 1996). Fatty acids can also be distributed into monounsaturated fatty acids which comprise only 1 double bond and polyunsaturated fatty acids which comprise 2 or more than 2 double bonds (Weiss, 1983). Normally, unsaturated fats have very low melting points and exist in liquid form at room temperature (Marangoni and Narine, 2002; O'Keefe, 2002) (Figure 2.1).

2.3.2. Physical Properties

2.3.2.1. Crystal Formation

Fat molecules can normally bind one another to form crystals. When the liquid fat is cooled down, the molecular movement is decreased; the molecules will usually be attracted to one another by Van der Waals forces. These forces are of minor significance and weak in small molecules, however, the effect is collective in long-chain or large molecules (Clements, 1971; O'Keefe, 2002).

Crystal formation is dependent on the symmetry of the fat molecules and the similarity of the fatty acid chain length (Marangoni and Narine, 2002). Fats comprising asymmetrical molecules and molecules comprising the coil structure of double bonds line up in a very difficult manner since they can't pack together very closely in space. Symmetrical fat molecules that line up easily tend to create large crystals since they possess high melting points consequential from less energy eliminated before they crystallize. Conversely, asymmetrical molecules incline to produce small crystals since more energy should be eliminated before they crystallize (Goldfine, 1968; Wnson et al., 1997).

2.3.2.2. Polymorphism

Fat polymorphism is when the fat can occur in diverse crystalline forms. A fat might crystallize in one of four different crystal forms, dependent on conditions during the crystallization and the composition of the fat. The least stable and smallest crystals are known as a-crystals. They are created when the fats are chilled quickly. Due to instability, they change eagerly to b^0-crystals, which are very small needle-like crystals about one mm long. Fats that can create stable b-crystals are very good for utilization as shortenings since they can be easily creamed and give a very smooth texture. Unstable b^0-crystals alter to intermediate crystal form, almost 3 to 5 mm in size, and lastly transform to coarse beta b-crystals, which normally range from 25–100 mm in length (Timms, 1991; Vaclavik and Christian, 2014). These large crystals possess the highest melting point.

The creation of small crystals is normally favored by quick cooling with agitation. This permits the formation of numerous small crystals, in place of the slow growth of the fewer large crystals (Akoh and Min, 2002).

The formation of large or small crystals depends on the homogeneity and cooling rate of fats (Marangoni and Narine, 2002). The growth of the large crystals takes place if the cooling rate is slow and vice versa. The more the fat is heterogeneous, the more possibility that the molecules create small stable crystals. Large crystals are created by homogeneous fats and the small crystals are created by heterogeneous fats. Increasing heterogeneity of the fatty acids composition averts the creation of large crystals. Fats with quite small crystals are usually harder fats, possess smooth fine texture, and seem to be less oily since the oil is existent as the fine film surrounding crystals, while the reverse is also true of fats with the large crystals. Smaller crystals are anticipated if the fat adds aeration to a food.

2.3.2.3. Melting Points

The melting point is generally an index of a force of attraction among the molecules. A strong attractive force designates a good degree of fitness between the molecules. Molecules that don't fit together well don't have strong attractive forces binding them together, and thus they possess lower melting points (Marangoni and Narine, 2002).

The melting points range of oil or fat is based on the constituent of the fatty acids (Akoh and Min, 2002). The melting point of oil or fat is truly a range, not the sharply defined temperature. Each oil or fat comprises triglycerides that normally melt at different temperatures, dependent on their constituent fatty acids. The more heterogeneous fatty acids will normally fit together less easily; therefore they will usually have lower melting points. Some fats have an extensive melting range, while others, like chocolate or butter, have a narrow melting range. Chocolate has a narrow melting range that is very close to the temperature of the body, and this is responsible for its specific melt in your mouth property.

The melting point is concluded by the length of the fatty acid chain, isomeric configuration, and a number of the double bonds. The first point is the length of the fatty acids chain: fatty acids with long chains possess higher melting point than the fatty acids having short chains, as there is more probability for attraction amongst long chains as compared to short chains; for example, butyric acid has the melting point of 18 Fahrenheit, while stearic acid has the higher melting point of around 157 Fahrenheit.

The second factor on the melting point is the isomeric configuration. The cis double-bonds configuration presents a number of the coil structures which are very problematic to pack together closely into the molecule as compared to the trans configuration, therefore it has a lower melting point (Weiss, 1983). The third influence is a number of double bonds: with a higher number of double bonds, lower will be the melting point.

2.3.3. Chemical Reaction

2.3.3.1. Hydrolytic Reaction

Off-flavors in the products based on fats are triggered by hydrolysis of the fatty acids from triacylglycerols (Akoh and Min, 2002). Free fatty acids might produce off-flavor, decrease oxidative stability, trigger foaming, and decrease smoke point. In hydrolytic rancidity, the free fatty acids are freed from glycerol backbone consequential in the development of the off-flavors. Free fatty acids yield off-aromas and fatty acids with long chains produce soapy tastes (Min and Boff, 2002). Triacylglycerol hydrolysis also takes place in frying oils during the high-temperature processing, water from the fried food encouraging hydrolysis of the triacylglycerol. As the content of free fatty acid of the frying oil rises, oxidative stability and smoke point decreases and the affinity for foaming increases. Triacylglycerol hydrolysis will take place at extreme values of pH. However, free fatty acids with short chains occasionally contribute to anticipated flavor profiles, like in cheese products or yogurt.

Enzyme lipases generally play a vital role in the liberation of free fatty acid. Enzyme lipases existent in living tissues of the raw food materials and the activity can take place naturally like in the processing of olive oil. Oils from consequent pressing and the oil extracted from pomace have a higher concentration of free fatty acid as the cellular matrix is disrupted further giving ample time to lipases to hydrolyze triacylglycerols (Rousseau and Marangoni, 2002).

2.3.3.2. Oxidative Rancidity

This is the main type of rancidity. In this procedure, the unsaturated fatty acid is subjected to autoxidation. The more the number of double bonds existent, the greater the chance for the addition of oxygen (O_2) to double bonds, escalating the danger that the oil or fat will become rancid (Zhuang et al., 2002).Autoxidation is generally an oxidative intricate reaction and is stimulated by heat, certain metals (copper and iron), light, and enzymes called the lipoxygenases (Min and Boff, 2002). There are three stages in the process of autoxidation reaction:

(a) Initiation;
(b) Propagation; and
(c) Termination.

The first stage of this reaction involves the creation of free radicals. The free radical outcomes if the hydrogen on C atom nearby one carrying the double bond are evacuated (Buettner, 1993). This reaction is disseminated which moves another hydrogen from a new unsaturated fatty acid, creating another free radical. The reaction repeats and the stage is known as the propagation stage that produces activated peroxide. The hydrogen liberated then unites with peroxide to form the hydroperoxide. Hydroperoxides are quite unstable and usually decompose into compounds having shorter C chains, like volatile fatty acids, ketones, and aldehydes (Decker and McClements, 2001). These are accountable for the characteristic aroma of rancid oils and fats. The third and final stage of this reaction comprises the reaction of the free radicals to create non-radical products. Removal of all the free radicals is the only method to end the oxidation reaction.

Oxidation can be delayed or prevented by escaping circumstances that would assist as the catalysts for reaction (e.g., presence of light, water vapor, and oxygen, high temperature) and addition of the antioxidant (e.g., tocopherols, EDTA, BHT (butylated hydroxytoluene), BHA (butylated hydroxyanisole), propyl gallate, and TBHQ (tertiary-butyl hydroquinone)) (Buettner, 1993).

2.3.4. Functional Properties in Food Products

The functionality of the triacylglycerol plays a significant role in determining the sensory and physical properties of the food products, like texture, flavor, and appearance.

2.3.4.1. Texture

The functionality of triacylglycerol's in the food texture is decided by the physical state of lipids in the food matrix. For partly crystalline fats, like chocolate, shortenings, butter, margarine, and baked products, the texture is primarily determined by the morphology, interactions, and concentration of the fat crystals (McGrady, 1994; Pszczola, 2000). Especially, the melting profile of fat crystals generally plays the main role in concluding properties like texture, spreadability, mouthfeel, and stability (Kinsella, 1981; James, 2009).

The concentration and behavior of lipid in the foods matrix will also add to the whole texture; for instance, oil in water food emulsion is decided by the existence of fat droplets (e.g., desserts, dressings, creams, and mayonnaise). In these kinds of systems, the texture is signified by the viscosity of the entire system. In the water in oil food emulsions, the texture is characterized by the complete rheology of the system, which is determined largely by the rheology of oil as a leading phase. The rheology of the products is decided by the interactions and morphology of the fat crystals existent during storage and crystallization conditions. The formation of a 3-D network of combined fat crystals in the continuous phase delivers fat products with spreadability (Gontard et al., 1994; Moran, 1994). Conversely, the interaction of the lipids with some other compounds in the food influences their texture by creating a network of interrelating fat crystals that provide characteristics of snap or firmness (Marangoni and Narine, 2002).

The water in oil emulsion comprises of water droplets implanted in the lipid phase that comprises a network of combined crystals. The fat crystals made are primarily in a polymorphic form but are normally transformed to the stable b^0 polymorphic arrangement during the crystallization. It is vital to control the amount of this conversion during the process of manufacturing as this regulates the strength and number of the bonds made amongst the fat crystals and thus the rheology of the final product. Conversely, it is also significant to avert the polymorphic conversion from the b^0 to more stable b arrangement during storage, as this leads to the creation of large crystals that are observed as grainy or gritty in the mouth. This conversion can frequently be prohibited by adding the surfactants that hinder with the polymorphic conversion, selecting lipids that won't make b crystals, or by a suitable blending of lipids to support the b^0-crystal habit (Lawler and Dimick, 2002).

2.3.4.2. Appearance

The typical appearance of several food products is characterized by visual observation like color and opacity. The color of unpackaged oils, like salad or cooking oils, is primarily determined by the existence of pigment contaminations that adsorb light, like chlorophyll and carotenoids. The opacity of fat is dependent on the concentration, shape, and size of the fat crystals existent. Solid fats are normally optically opaque due to scattering of the light by fat crystals existent, while liquid oils are normally optically clear. Food emulsions generally appear optically opaque as the light passing through these emulsions is dispersed by the droplets (McClements, 2002).

Instability problems linked with the crystallization of fat in the products based on lipid-like chocolate can trigger bloom. Bloom demonstrates itself as the large grayish look on the surface of the product (Lonchampt and Hartel, 2004). The fluctuating temperature during the storage triggers the fats phase to melt and then recrystallize (Erickson and Frey, 1994). Bloom can frequently be prevented or retarded by utilizing surfactants that restrict the crystal conversion or by carefully governing the storage temperature to avert fat polymorphic phase conversions (Erickson and Frey, 1994; Lonchampt and Hartel, 2004).

2.3.4.3. Flavor

Edible oils and fats add to the flavors of food. Different composition of fatty acid in oils and fats has characteristic flavor profiles. The flavor of several food products is influenced indirectly by the lipid phase as flavor compounds can divide between oil, gaseous, and water regions inside the food matrix conferring to their volatilities and polarities (McClements, 2004). Due to this reason, the apparent taste and aroma of foods are frequently strongly impacted by the concentration and type of lipids present. Physical states of the lipids also impact the mouthfeel of several food products (Walstra, 2003), for instance, oily mouthfeel yielded by liquid oils and grainy or sandy mouth feel made by fat crystals (Moran, 1994).

2.4. PROTEINS

Proteins are existent in all of the living things and play a vital role in numerous biological processes. The nourishing energy value of the proteins is around 4 kcal/g. Amino acids are needed in building blocks for protein biosynthesis. Damodaran (2008) described food proteins like those that are effortlessly digestible, nutritionally adequate, nontoxic, functionally helpful in food products, sustainable agriculturally, and available in abundance. Proteins add to the flavors of food, predecessors for odor compounds and colors made during the processing of food. Proteins have the capability to stabilize or build gels, foams, fibrillar structures, and emulsions, which is crucial in certain food products (Christakos et al., 1989; Uniacke-Lowe et al., 2010). The proteins can normally be attained from milk, grain, oilseeds, meat, legumes, algae, bacteria, and yeasts.

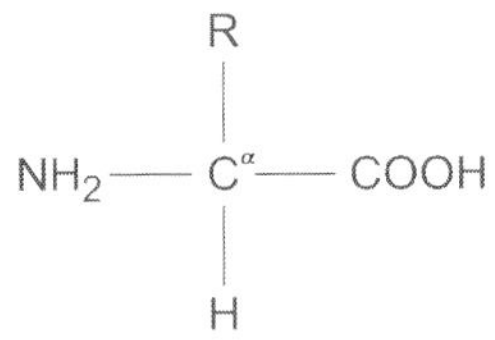

Figure 2.2: The general structure of the amino acid.

Source: https://link.springer.com/referenceworkentry/10.1007%2F978-3-642-36605-5_35.

Table 2.4: Common Amino Acids Discovered In Proteins

Type	Name	Abbreviation (three and one letter symbols	
Simple	Glycine	Gly	G
Aliphatic	Leucine	Leu	L
	Alanine	Ala	A
	Valine	Val	V
	Isoleucine	Ile	I
Heterocyclic	Tryptophan	Trp	W
	Proline	Pro	P
	Histidine	His	H
Aromatic	Tyrosine	Tyr	Y
	Phenylalanine	Phe	F
Hydroxy	Threonine	Thr	T
	Serine	Ser	S
Acidic	Glutamic acid	Glu	E
	Aspartic acid	Asp	D
Amide	Glutamine	Gln	Q
	Asparagine	Asn	N
Basic	Arginine	Arg	R
	Lysine	Lys	K
Sulfur-containing	Methionine	Met	M
	Cysteine	Cys	C

2.4.1. Chemical Structure and Cataloging

Twenty amino acids are creating the building blocks of numerous proteins. Every amino acid comprises the carboxylic acid group and a primary amine, which is displayed in Figure 2.2. They tie by peptide bonds created between a-carboxylic acid and a-amino groups of adjacent amino acids in the polypeptide sequence. The amide association in the proteins is a fractional double bond (Wieser, 2007; Masters and Friedman, 1980).

In the simplest example, amino acetic acid or glycine (R = H). In the other amino acids, R is aromatic, heterocyclic, or an aliphatic residue and might integrate other functional groups displayed in Table 2.4.

Several side chains of amino acids are included in intra- and intermolecular interfaces in proteins and categorized into several methods, which are exhibited in Table 2.5. Proteins are macromolecules and their name was originated from the Greek word "proteois," meaning first of the kind. Proteins contain 50–55% C, 6–7% hydrogen, 20–23% oxygen, 12–19% nitrogen, and 0.2–3.0% sulfur on the wt/wt basis. Protein synthesis takes place in ribosomes and then consequently some of the proteins get altered by cytoplasmic enzymes (Damodaran, 2008; Belitz et al., 2009).

Proteins reformed by an enzyme or made intricate with nonprotein constituents (prosthetic groups) are known as conjugated (hetero-proteins) proteins while the modified proteins are called homoproteins. Numerous proteins like glycoproteins, lipoproteins, metalloproteins, nucleoproteins, and phosphoproteins are examples of the conjugated proteins. Additionally, based on the structural arrangement, proteins can generally be classified as fibrous and globular proteins. Globular proteins prevail in spherical or ellipsoidal shapes because of the folding of the polypeptide chain on itself whereas fibrous proteins are the rod-shaped molecules comprising twisted linear polypeptide chains (Cecil and McPhee, 1959; Chalker et al., 2009). Enzymes are the globular proteins while collagen, fibrin, and keratin are fibrous proteins. Based on the biological functions, proteins can normally be classified as structural proteins, transfer proteins, contractile proteins, enzyme catalysts, protective proteins, hormones, storage proteins, and antibodies (Kinsella and Melachouris, 1976; Damodaran, 2008;).

Table 2.5: The Base for the Organization of Amino Acids

Basis for classification	**Amino acids**
Charged side chains	Arginine, glutamic acid, aspartic acid, histidine, and lysine
Uncharged, polar side chains	Serine, threonine tyrosine, asparagine, cysteine, and glutamine
Nonpolar, uncharged side chains	Methionine, glycine, valine, leucine, isoleucine, alanine, phenylalanine, proline, and tryptophan
Nonessential amino acids	Glycine, alanine, serine, cysteine, tyrosine, proline, asparagine, aspartic acid, glutamine, and glutamic acid
Essential amino acids	Valine, leucine, phenylalanine, tryptophan, isoleucine, methionine, histidine, lysine, threonine, and arginine

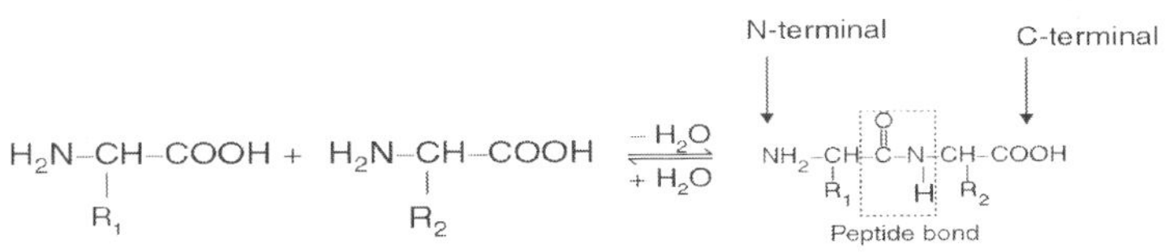

Figure 2.3: Peptide bond with C- and N-terminal.

Source: https://link.springer.com/referenceworkentry/10.1007%2F978-3-642-36605-5_35.

2.4.2. Structure of Proteins

Amino acids are associated covalently via the a-carboxyl group of an amino acid and a-amino group of another amino acid via a peptide or amide bond to form peptides, polypeptides, oligopeptides, and proteins (Ustunol, 2015). Peptides are represented by a number of the amino acid deposits as tri-, di-, tetrapeptides, etc. and the term oligopeptides is utilized for those with ten or fewer amino acid remainders. Peptides with higher molecular weight (MW) are called polypeptides. The transition of a polypeptide to protein is rather vague, but the limit is usually supposed to be at the MW of around 10 kd; i.e., around 100 residues of amino acids are required in the chain for the polypeptide to be known as protein. The residue of amino acid having a free amino group is placed on the left side at all times. The amino acids at the ends of the chain are represented as C-terminal and N-terminal amino acid residues (Belitz et al., 2009), which are displayed in Figure 2.3.

Proteins are made from amino acids via amide linkages. The structure of protein depends on the sequence of an amino acid (the primary structure), which concludes the molecular conformation. Proteins sometimes take place as aggregates of molecules which are usually arranged in an organized geometric fashion (quaternary structure) (Belitz et al., 2009). Therefore proteins possess primary, secondary, quaternary, and tertiary structures, which are displayed in Figure 2.4.

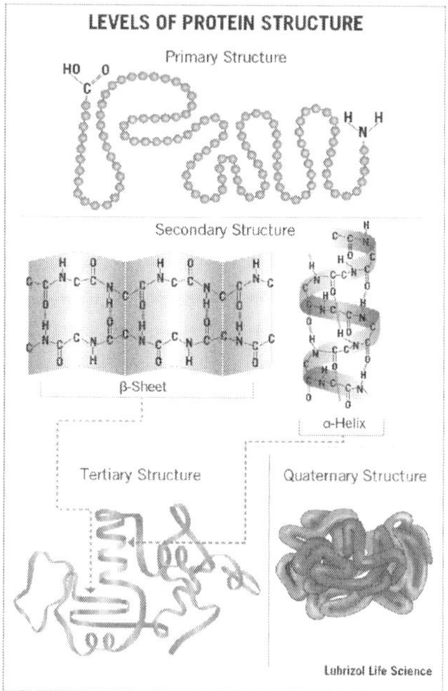

Figure 2.4: Different levels of the protein structures.

Source: www.particlesciences.com.

2.4.2.1. Primary Structure

It relates to the linear series in which the essential amino acids are linked covalently via amide bonds, also called the peptide bonds. The primary structure is decided by post-translational covalent modifications and its genetic code. The primary structure of the protein concludes the functional and physicochemical properties of the food protein (Damodaran, 2008; Ustunol, 2015).

2.4.2.2. Secondary Structure

The primary structure provides the amino acid sequence in the protein chain whereas the secondary structure discloses the arrangement of a chain in the space. A secondary structure might result from periodic or periodic structures. In the periodic structures consecutive residues of amino acid possess diverse sets of dihedral angles while extended and helical structures are the periodic structures which outcome from consecutive residues of amino acid in the segment having a similar recurring set of c and j angles (Damodaran, 2008; Ustunol, 2015). The b-, 3_{10}-, and a-helices are helical structures, however, the right-handed a-helix is common and most stable in the food proteins. Proteins helical structures are created when the c and j angles of consecutive residues of an amino acid are twisted to the same set of values. Stability to the a-helix is given by hydrogen bonding of the C=O and N-H groups of fourth preceding residue. The bonds of hydrogen are positioned parallel to the axis of the helix. The a-helical structure discovered in proteins is usually amphiphilic. The proline, cyclic imino acid, can't form a-helices. Proline is thought to be the a-helix breaker. It can't form hydrogen bonds because of the deficiency of the hydrogen on the nitrogen atom. Proteins comprising high levels of residues of proline will usually take random periodic structures. An as_1-casein and b-casein possess about 8.5% and 17% proline of the entire amino acid, respectively. Proline residues are distributed uniformly, and a-helices aren't existent in these proteins (Damodaran, 2008; Ustunol, 2015).

2.4.2.3. Tertiary Structure

It relates to the spatial organization achieved when a chain of linear protein with the secondary structure segments generally folds further into the compact 3-D form. Formation of the tertiary structure includes hydrophobic, van der Waals, electrostatic, and hydrogen bonding amongst several groups of proteins. The folding of the protein into the tertiary structure outlines the shape and size of that protein. Folding of proteins normally brings hydrophobic deposits at the interior along with charged and hydrophilic remainders on the surface of a molecule (Damodaran, 2008; Ustunol, 2015). The presence of hydrophobic and hydrophilic residues in polypeptide concludes the shape of proteins. Globular proteins are ample in the hydrophobic residues whereas rod-shaped or elongated proteins are ample in evenly distributed hydrophilic residues (Kinsella, 1981, 1982; Nelson and Cox, 2013).

2.4.2.4. Quaternary Structure

It relates to the spatial organization of the protein when it comprises more than one chain of a polypeptide, which is also known as oligomer or subunit. The formation of the quaternary structure is mainly driven by a thermodynamic necessity to bury unprotected hydrophobic surfaces of the subunits. The quaternary structure is stabilized further by non-covalent interactions like hydrogen bonding and electrostatic and hydrophobic interactions (Damodaran, 2008; Ustunol, 2015).

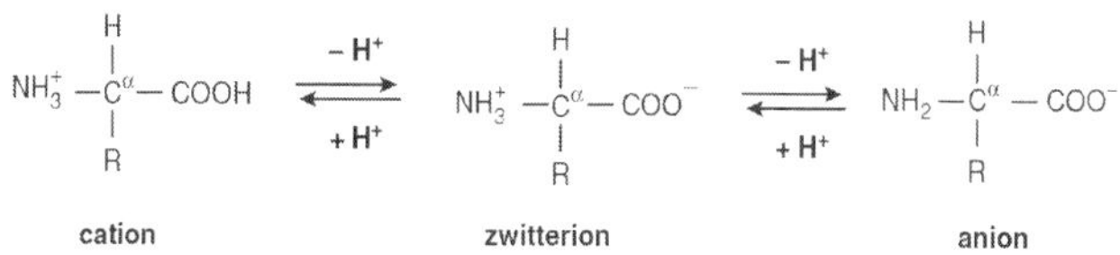

Figure 2.5: Separation of amino acids.

Source: https://link.springer.com/referenceworkentry/10.1007%2F978-3-642-36605-5_35.

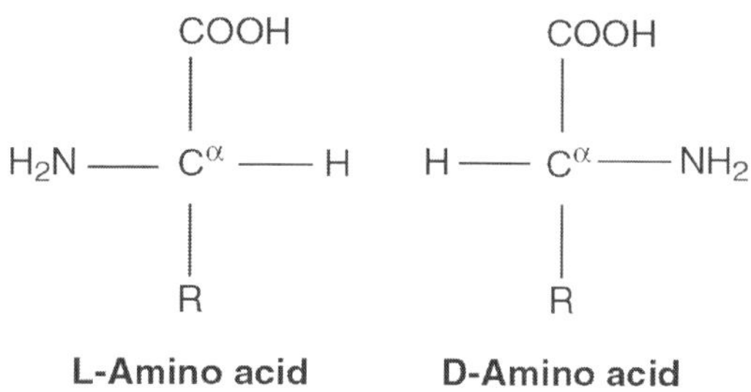

Figure 2.6: L- and D- amino acids.

Source: https://link.springer.com/referenceworkentry/10.1007%2F978-3-642-36605-5_35.

Folding the random chain of the polypeptide into its 3-D structure is a very complex process, and several forces are involved. These forces might be grouped into two categories below:

(i) Intermolecular interactions outcome from the surrounding solvent (i.e., electrostatic, and hydrophobic interactions, hydrogen bonding); and

(ii) Intramolecular interactions outcome from the forces intrinsic to a protein molecule (i.e., stearic, van der Waals) (Damodaran, 2008; Ustunol, 2015).

2.4.3. Physical Properties

2.4.3.1. Acid-Base/Dissociation/Properties

Amino acids are normally present in the aqueous solution, depending on the pH, as zwitterions, anions, or cations which are displayed in Figure 2.5. The acidity of the amino acid of the carboxyl group (COOH) is usually higher and the basicity of the amino group is lower. Amino acids act as both bases and acids, i.e., they are ampholytes. At almost neutral pH, both the a-carboxyl and a-amino groups are ionized and the molecule is a zwitterion or a dipolar. The pH at which this dipolar ion is neutral electrically is known as the isoelectric point (pI) (Damodaran, 2008; Belitz et al., 2009).

2.4.3.2. Optical Activity and Configuration Properties

Amino acids, excluding glycine, have one chiral center as a minimum and, therefore, are optically active. Isoleucine, 4-hydroxyproline, and threonine have two asymmetric C-atoms, therefore each has four isomers (Belitz et al., 2009). The D- and L-amino acids are displayed in Figure 2.6.

All proteins discovered in nature comprise only L-amino acids. D-amino acids do occur in nature, for instance, in numerous peptides of the microbial origin. The particular rotation of the amino acids in an aqueous solution is influenced strongly by pH. It passes through the minimum in neutral pH range and increases after the addition of bases or acids. The optical activity of the proteins isn't only due to the asymmetry of the amino acids but also due to the chirality occasioning from the arrangement of the peptide chain (Belitz et al., 2009).

2.4.3.3. Solubility

The solubility of the amino acids in water (H_2O) is highly variable. Proline is enormously soluble; hydroxyproline, alanine, and glycine are quite soluble, while tyrosine (Tyr) and cystine have low solubility's in the water. The addition of bases or acids enhances the solubility via salt formation. The amount of solubility of the amino acids in hydrolysates of proteins is different than that noticed for the individual constituents (Wilding et al., 1984; Panyam and Kilara, 1996). The solubility in the organic solvents isn't very good due to the polar characteristics of amino acids. All of the amino acids are insoluble in the ether. Protein solubility is inconstant and is impacted by the number of apolar and polar groups and their organization along with the molecule (Chou and Morr, 1979; Belitz et al., 2009).

2.4.3.4. UV absorption

Aromatic amino acids like phenylalanine, tryptophan (Trp), and Tyr absorb in the ultraviolet (UV) range of spectrum with the absorption maxima at around 200 to 230 and 250 to 290 nm. Trp and Tyr also show fluorescence in the UV region (250–300 nm). Histidine, methionine, and cysteine absorb between 200–210 nm. Absorption readings at the 280 nm are utilized for the determination of peptides and proteins (Damodaran, 2008; Belitz et al., 2009).

2.4.3.5. Sensory Properties

While the quality of taste of the amino acids is dependent on peptides, configuration, excluding sweet dipeptide esters of the aspartic acid, is bitter or neutral in taste with no association to configuration. The intensity of taste is impacted by the hydrophobicity of side chains. The taste intensity doesn't appear to depend on the sequence of amino acids. Bitter-tasting peptides can also occur in the food products after proteolytic reactions (Jambrak et al., 2008; Belitz et al., 2009).

2.4.4. Chemical Reactivity of the Amino Acids

Amino acids exhibit the common reactions of both amines and carboxylic acids. Reaction specificity is because of the existence of both amino and (COOH) and some other reactive groups like amino, carboxyl, phenolic, hydroxyl, sulfhydryl, thioether (Met), guanyl, and imidazole in free proteins and amino acids. Reactions happening at 100–220°C, like in cooking, baking, and frying, are particularly appropriate to food chemistry (Damodaran, 2008; Belitz et al.2009). Numerous type of reactions which involve (i) amino groups (e.g., guanidation, arylation, reductive alkylation, succinylation, acetylation, thiolation, deamination), (ii) carboxyl group (e.g., reduction, decarboxylation, esterification,), (iii) sulfhydryl group (blocking, oxidation), (iv) threonine and serine (e.g., esterification), and (v) methionine (e.g., b-propiolactone, alkyl halides,) take place. Some of these reactions can be utilized to change the hydrophobic and hydrophilic properties and functional properties of peptides and proteins along with quantifying amino acids and particular residues of amino acid in proteins (Damodaran, 2008; Belitz et al., 2009).

2.4.5. Functional Properties in Food Products

Proteins have a range of vibrant functional properties and signify the most vital class of the functional ingredients. The functionality of the food proteins refers to chemical and physical properties that impact the role of proteins in the food systems during processing, preparation, storage, and consumption (Damodaran, 2008). Functional properties of the protein are dependent on physicochemical properties such as shape, size, sequence, and composition of amino acid, structural levels of proteins, distribution of charges, molecular flexibility, hydrophobicity/hydrophilicity ratio, and ability to interact/react with other components (such as lipids, sugars, salts, polysaccharides, and minor components). Numerous functional properties such as organoleptic (odor, flavor, color), hydration (gelling, syneresis, thickening, solubility, swelling, wettability, viscosity, gelation, water sorption), kinesthetic (texture, mouthfeel, grittiness, smoothness), rheological/textural (cohesiveness, chewiness, aggregation, elasticity, adhesiveness, dough formation, network formation, extrudability, texturizability), and surface (emulsification, foaming, film formation) are usually possessed by proteins. Proteins exhibit versatility during the processing. They can form structures and networks and provide necessary amino acids. Moreover, they interrelate with other components and increase the quality attributes of the foods (Singh, 1991; Kinsella et al., 1994).

2.4.5.1. Solubility of Proteins

The solubility of proteins is variable. Protein solubility is thought of as that ratio of nitrogen in the protein product which under specific conditions is generally in the soluble state. Proteins can be completely or partly soluble or entirely insoluble in water. Protein solubility offers helpful information on foams, gels, and emulsions. Protein solubility is impacted by the composition and sequence of amino acid, conformation, and MW and content of non-polar and polar groups in the amino acids. Water molecules bind to backbone peptide groups, charged groups, amide groups of asparagines and glutamine, hydroxyl groups of serine, Tyr, and threonine residues and non-polar residues in proteins. Ionic strength, pH, temperature, type of solvent, and processing conditions disturb the solubility of proteins. Solubility is augmented if electrostatic repulsion amongst the molecules is higher as compared to the hydrophobic interactions. At the pI (isoelectric point), proteins possess a net 0 charge, attractive forces dominate, and molecules incline to associate, consequential in insolubility. Above the pI, the net charge is usually negative and solubility of proteins is enhanced. Protein-water interfaces increase at the values of pH lower or higher than the pI because protein carries a negative or positive charge.

2.4.5.2. Water Holding Capability of Proteins

Water hydration and holding, water binding, water retention, water absorption, and water imbibing are some terms utilized interchangeably for the collaboration of proteins with water. The WHC of foods can normally be defined as the capability to hold the added and its water during the stage of application of forces, centrifugation, pressing, or heating. WHC plays an important role in the creation of food texture, particularly in baked doughs and comminuted meat products. Swelling is a vital functional property of proteins as it is the chief step in the solvation of proteins and can be described as an impulsive uptake of water by the protein matrix. Water retention is an important factor in the functionality of proteins as it affects the color, texture, and sensory properties of the products. Water binding is dependent on the conformation and composition of the protein molecules. Water intermingles with proteins in several ways, and substantial amounts of water restricted by proteins are engaged by hydrogen bonding. Interactions between hydrophilic groups of protein side chains and molecules of water occur through hydrogen bonding. Structural water is usually held by the hydrogen bonding amongst polypeptide groups of proteins. Binding of water to the proteins is linked to polar hydrophilic groups, like imino, carboxyl, hydroxyl, carbonyl, amino, and sulfhydryl groups. The ability of proteins to preserve moisture is disturbed by the number and type of the polar groups in the protein polypeptide chain. Binding of the water is because of the dipolar character of water. Proteins that contain several charged amino acids will incline to bind large quantities of water. Water binding of the proteins can be anticipated from the amino acid composition. The absorbed water is bound tightly to the molecules of protein. Water binding of the protein is impacted by protein concentration; ionic strength; temperature; pH; the presence of some other components of foods like hydrophilic polysaccharides, salts, and lipids; length and rate of heat treatment; and the conditions of storage (Zayas, 1997; Belitz et al., 2009).

2.4.5.3. Foam Formation and Stabilization

Proteins function as foam-stabilizing and foam-forming components in several foods, e.g., baked goods, desserts, sweets, and beer. The most widely utilized protein foaming agents are gelatins, casein, egg white, other milk proteins, gluten, and soy proteins. Serum albumin foams quite well, whereas egg albumin doesn't. Foams comprise of an aqueous constant phase and a gaseous dispersed phase. Foam can be stated as a 2-phase system comprising of air cells isolated by the thin continuous liquid layer known as a lamellar phase. Proteins stabilize by forming stretchy, cohesive films around the gas bubbles. Protein is adsorbed at the interface through hydrophobic areas trailed by partial unfolding. Protein adsorption assists in the creation of new interfaces by decreasing surface tension. The more rapidly a molecule of protein diffuses into the interfaces and the easily the molecule is denatured there, the more the protein can foam. Foam stability denotes the capacity of proteins to stabilize foam against mechanical and gravitational stresses. Foams collapse as large gas bubbles matures at the cost of smaller bubbles. The protein films counter this disproportionation. An ideal foam-stabilizing and foam-forming protein are categorized by the low MW, good solubility, high surface hydrophobicity, a small net charge, and easy denaturability (Zayas, 1997; Belitz et al., 2009).

2.4.5.4. Gel Formation

A gel is a midway phase between a liquid and a solid. A gel is a considerably diluted system that shows no steady-state flow. Gelation is a fundamental procedure in the processing of several foods, comminuted meat and fish products, milk gels, other meat products, bread doughs, fruit jellies, coagulated egg white, pie, and cake fillings, and others. Protein gels might be used to simulate the mouthfeel and textural properties of lipids. Gels are made when partly unfolded proteins grow uncoiled polypeptide segments that intermingle at particular points to form a 3-D cross-linked network. The partial unfolding of the proteins with minor changes in the secondary structure is needed for gelation (Zayas, 1997; Belitz et al., 2009).

2.4.5.5. Emulsifying Effect

An emulsion is the disperse system of one or more than one immiscible liquids. They are normally stabilized by the emulsifiers-compounds that produce interface films and therefore avoid the disperse phases from flowing together. The adsorption of protein at the boundary of an oil droplet is favored thermodynamically since the hydrophobic residues of amino acid can then avert the hydrogen bridge network of surrounding water molecules. Additionally, contact of the proteins with an oil droplet outcomes in the displacement of the water molecules from hydrophobic regions of an oil-water (o/w) boundary layer. Thus, the protein suitability as an emulsifier is dependent on the rate at which the protein diffuses into the boundary and on the deformability of the conformation under the impact of interfacial tension. The rate of diffusion is dependent on the temperature and MW, which can be impacted by the ionic strength and pH. A protein having supreme potentials as an emulsifier for o/w (oil-in-water) emulsion would have a comparatively low MW; balanced composition of amino acid in terms of polar, nonpolar, and charged residues; well-developed surface hydrophobicity; good water solubility; and a comparatively stable conformation. The b-casein molecule fulfills these requirements due to the less pronounced secondary structures and no cross-links because of the lack of SH groups (Belitz et al., 2009).

2.5. CARBOHYDRATES

2.5.1. Chemical Structure and Cataloging

Carbohydrate generally has an elemental composition in nature, namely, $Cx(H_2O)y$, which indicates molecules comprising C atoms along with oxygen and hydrogen atoms in a similar ratio as they exist in water (Fennema, 1996). However, the majority of carbohydrate compounds yielded by living organisms don't have a simple empirical formula (Vaclavik and Christian, 2014). Instead, most of the natural carbohydrates are categorized as simple sugars, oligomers, and polymers. However, this chapter starts with a presentation of the simple sugars and from there builds to larger and more intricate structures (Whistler, 1964; Capon, 1969).

2.5.1.1. Monosaccharides

Monosaccharides are very simple carbohydrates comprising 3–8 C atoms, but only the ones with 5–6 C atoms are common. Two of the most significant ones in the foods are 6-carbon sugars, fructose, and glucose which possess the general formula $C_6H_{12}O_6$ (Fennema, 1996; El Seoud et al., 2007).

Glucose is classified as an aldose sugar because of an aldehyde group situated on the first C atom of a chain. It is traditional to number the C atoms along the chains hence that the C atom having the highest number is far away from the aldehyde group. The aldehyde group thus is situated on C atom one in the glucose (Belitz et al., 2009; Vaclavik and Christian, 2014).

In nature, two isomers of glucose prevail, which are the mirror images of one another: L-glucose and D-glucose. D-glucose is normally the isomer that exists naturally. There are two series of aldose sugars, called the l-series and d-series, in which every isomer is made by adding the CHOH groups to build the chain of C, beginning from the tiniest aldose sugar, which is L- or D-glyceraldehyde (Tvaroŝka and Bleha, 1989; Cho et al., 1999).

As displayed in Figure 2.7, every H-C-OH group inside the chain is asymmetrical. The highest-numbered asymmetric C atom of every d-series sugar possesses a similar configuration as D-glyceraldehyde, instead of its l-isomer. The highest-numbered asymmetric C atom in glucose is carbon-5, which is designated the reference C atom because of structure configuration fortitude if the sugar is called as the l series or d series. The hydroxyl group fixed to it is known as a reference hydroxyl group, which is on the right side in the d-series sugar (Stick and Williams, 2009).

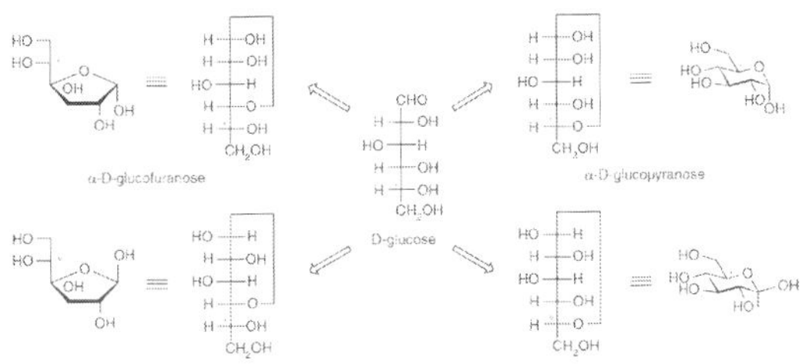

β-D-glucofuranose β-D-glucopyranose

Figure 2.7: The chief isomers of D-glucose (Fischer projections) (Reprinted with authorization from Withers (2015)). Anomeric center in CAZypedia.

Source: http://www.cazypedia.org/index.php/Anomeric_centre_ (alpha_ and_beta).

2.5.1.2. Oligosaccharides

The decreasing group of two monosaccharides can link to one of hydroxyl (OH) groups on another via glycosidic bond, to produce disaccharides. More links of the glycosidic bonds will yield to trisaccharides, tetrasaccharides, etc., characterized as oligosaccharides and eventually polysaccharides. The glycosidic bond is an outcome of condensation of the hydroxyl (OH) of the hemiacetal group of C-1 with hydroxyl (OH) of alcohol. When the glycosidic linkage is developed only amongst the lactol groups of two monosaccharides, then a non-reducing disaccharide is made, and when one lactol group and one alcoholic HO group are included, a reducing disaccharide outcome (Linhardt, 2003; Belitz et al., 2009).

2.5.1.3. Polysaccharides

When oligosaccharides possess more than 10 monosaccharides the unit is called the polysaccharide. Polysaccharides comprise units of monosaccharide bound to one another by glycosidic linkages. Their entire acidic hydrolysis produces monosaccharides. Partial enzymatic and chemical hydrolysis giving rise to polysaccharides is significant for structural elucidation, the investigation of which explains monosaccharide sequences and the types and positions of linkages (Stephen, 1995; Whistler and BeMiller, 1993).

Polysaccharides can only have a kind of sugar structural unit or various kinds of sugar units. Polysaccharides might have a linear pattern or a branched fashion of the monosaccharide unit linkage. The occurrence of branching sites and the length of the side chains can change greatly. The sequence of monosaccharide residue might be periodic, one period comprising one or various alternating structural units (amylose, cellulose, or hyaluronic acid), the sequence might contain longer or shorter segments with residues arranged periodically and separated by non-periodic segments, or the sequence might be non-periodic all over the chain. The conformation of monosaccharide structural units and the types and positions of the linkages in the chain decides the chain conformation of the polysaccharide (Fennema, 1996; Stephen, 1995).

2.5.2. Physical Properties

Hygroscopicity of sugar in the crystallized form is the capacity of moisture uptake by the sugars. The hygroscopicity changes and is dependent on the isomers present, sugar structure, and sugar purity (Belitz et al., 2009).

Monosaccharides incline to be soluble in the polar solvent; they are usually also soluble to the small extent in ethanol but aren't soluble in the organic solvents like benzene, ether, and chloroform. When the crystal sugars are caked together, the solubility will tend to decrease. Sugar solution in the high concentrations is utilized in the baking industry to maintain food moisture. The solubility of oligo- and monosaccharides in the water is good. However, anomers might differ considerably in their solubility, as demonstrated by a- and b-lactose (De Man, 1999). The sensory and physical properties of oligosaccharides are normally covered by monosaccharides as oligosaccharides comprise of not more than ten sugar molecules.

While sugars make true solutions, polysaccharides make colloidal solutions and are, therefore, hard to purify. The polysaccharides are amorphous and tasteless. Polysaccharides can be categorized as:

(a) **Homoglycans:** These are single monosaccharides connected, for example, starch, glycogen, and cellulose;

(b) **Heteroglycans:** These are two or more different components, for example, hemicelluloses, pectins, mucilages, and resins; and

(c) **Conjugated** compounds comprised of saccharides, lipids, or proteins (Stephen, 1995; Whistler and BeMiller, 1993).

Isolated polysaccharides are used to a great amount in food processing, either in the native or altered form, like gel-setting or thickening agents (starch, pectin, alginate, guaran gum), stabilizers for dispersions and emulsions, coating substances to safeguard sensitive food from unwanted change, film-forming, and inert fillers to escalate the ratio of indigestible substances in the diet.

2.5.3. Chemical Reactions

Carbohydrates can execute wide chemical reactions with several other compounds as besides having hydroxyl (OH) groups, some sugars and monosaccharides also have the carbonyl groups accessible for a reaction. This section will emphasize on two most usual reactions of the carbohydrates in food processing, Caramelization, and Maillard.

2.5.3.1. Maillard Reaction

Maillard is classified as non-enzymatic browning and generally plays a significant role in enhancing the appearance and taste of the cooked food, especially toasting, roasting, and baking food. The compounds that add to the Maillard reaction were amino acids and sugars in water, which was reported first by Louis Camille Maillard in, 1912. The chemistry essential for the Maillard reaction is quite complex. It encompasses not only one reaction path but the whole network of several reactions. The original complete reaction scheme of Hodge (Figure 2.8) has been advanced and elaborated by the food technologists, so the understanding of reaction is progressing steadily.

In the initial stage of Maillard, the reducing sugar, such as glucose, reacts with the compound possessing the free amino group occasioning in the condensation product N-substituted glucosamine. This product will trigger to form the ARP (Amadori rearrangement product). The consequent degradation of ARP depends on the pH of the system. At pH value of 7 or below, it experiences primarily 1,2-enolization with the creation of furfural when the pentoses are entangled or HMF (hydroxymethylfurfural) when the hexoses are entangled. At pH greater than 7 the degradation of the Amadori compound is considered to involve primarily 2,3-enolization and creation of reductones, like 4-hydroxy-5-methyl-2,3-dihydrofuran-3-one (HMF^{one}), and a range of fission products, comprising acetol, diacetyl, and pyruvaldehyde.

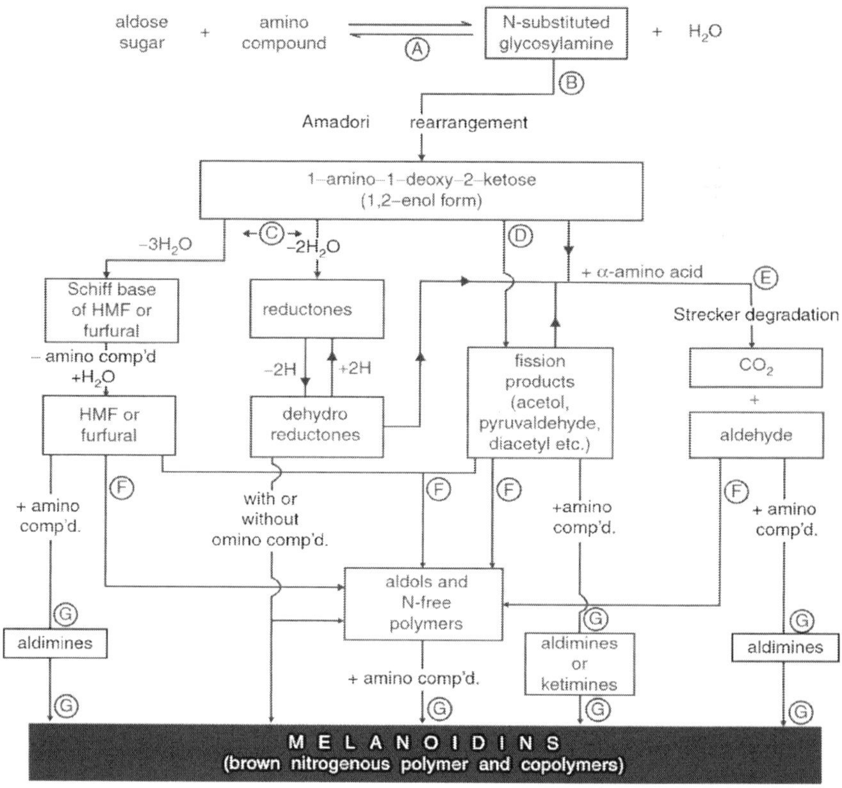

Figure 2.8: Maillard reaction scheme.

Source: https://link.springer.com/referenceworkentry/10.1007%2F978-3-642-36605-5_35.

All of these compounds are very reactive and indulge in further reactions. Carbonyl groups will normally condense with the free amino groups, which outcomes in the amalgamation of the nitrogen into reaction products. Dicarbonyl compounds will also react with the amino acids with the creation of a-aminoketones and aldehydes called the Strecker degradation. Consequently, in the progressive stage, a variety of reactions occur, including cyclizations, retroaldolizations, dehydrations, isomerizations, rearrangements, and further condensation. In the last stage, melanoidins were created, which are the brown nitrogenous polymers and copolymers (Martins et al., 2001).

2.5.3.2. Caramelization

Although it yields a similar brown color, caramelization is very different from Maillard. In caramelization, the decomposition of sugar occurs at very high temperatures and yields brown color; this reaction doesn't involve proteins and yields a different range of compounds than the Maillard, including aldehydes, organic acids, and ketones. The reaction is helped by the existence of a salt, acid, or base. Heating triggers dehydration of the molecule of sugar with the introduction of the double bonds or the creation of anhydrous rings. As in the Maillard browning, intermediates like furans and 3-deoxyosones are formed. The unsaturated rings might condense to form helpful, conjugated double-bond-comprising, brown-colored polymers. Catalysts increase the rate of reaction and are utilized to direct the reaction to particular types of caramel colors, acidities, and solubilities. Caramel served as flavor and colorant in cola, baked goods, acidic beverages syrups, dry seasoning, and candies (Damodaran et al., 2008).

2.5.4. Functional Properties in Food Products

In food, simple arrangements of carbohydrates have a major role as sweeteners in the candies and several other food products. The most common sensory property of sugars like glucose, sucrose, and fructose, is their sweetness, which changes depending on the particular sugar (Chinachoti, 1995; Murphy, 2001). Lactose, which is usually discovered as dominant sugars in the milk, is least sweet, while fructose is the sweetest among all. Moreover, sugars can make a solution as they are soluble in the water and make syrups with a high concentration in water. In the syrup production, water is usually evaporated to trigger crystal formation, therefore the molecular solutions are created due to hydrogen-bond interchange (Morris, 1990; Bhosale and Singhal, 2007). When the sugar is put in water, the water molecules instantaneously make hydrogen bonds with sugar molecules, therefore hydrating them and eliminating them from hydrogen bonds amongst the sugar crystals. At high concentrations, sugars perform as the food preservative which averts food spoilage by the development of microorganisms. This mechanism is linked to decreasing the water activity (aw) in food to the level below which bacterial development can't be supported. Examples of the foods conserved in this method include candied fruit and jams (Lee, 1983; Vaclavik and Christian, 2014).

Table 2.6: Applications of the Polysaccharides in Food

Area of application/food	Suitable polysaccharides
Stabilization of emulsions/ suspensions in chocolate milk and condensed milk	Carrageenan, algin, carboxymethylcellulose, pectin
Stabilization of ice cream against ice crystal melting, phase separation, formation; improvement of consistency	Algin, carrageenan, agar, gum tragacanth, gum arabic xanthan gum, guarana gum, modified starches, carboxymethyl cellulose, locust bean flour, methylcellulose
Stabilization of emulsions in low-fat kinds of margarine, coffee whiteners,	Carrageenan
Thickening and gelation of milk in the puddings made without and with heating, creams; enhancement of consistency	Pectin, algin, guaran gum, carrageenan, locust bean flour, modified starches, carboxymethyl cellulose
Water binding, enhancement of consistency, cream cheese, a yield increase of soft cheese, cheese preparations	Carrageen an, agar, karaya gum, guarana gum, gum tragacanth, locust bean flour, carboxymethyl cellulose, algin
Jellies for meat, vegetable, and fish products	Algin, carrageenan, agar
Water binding, stabilization of emulsions in the meat products	Agar, guaran gum, karaya gum, locust bean flour
Stabilization of protein foam in beer, meringues, whipped cream, chocolate marshmallows	Algin, carrageenan gum arabic, agar, karaya gum, xanthan gum

Stabilization and thickening, freeze-thaw stability of soups, prevention of synaeresis, sauces, mayonnaise, salad dressing, ketchup; obtaining body in low-fat and low-starch products	Gum tragacanth, algin, xanthan gum, guaran gum, karaya gum, locust bean flour, propylene glycol alginate, carboxymethyl cellulose, modified starches
Thickening and gelation of fruit pulp	Pectin, algin
Avoidance of starch retrogradation in cakes and bread, water binding in the dough	Agar, guaran gum, carrageenan, xanthan gum, locust bean flour
Sediment stabilization in the fruit juices, obtaining body in beverage powders	Algin, pectin, gum arabic, xanthan gum, propylene glycol alginate, guaran gum, methylcellulose
Gelation of jelly candies, glaze, icing, jelly beans, water desserts jellies	Pectin, algin, agar, gum arabic, carrageenan, modified starches
Stabilization of powdery odor emulsions, encapsulation of odor substances	–

Mono- and polysaccharides in suitable concentration or ratio back to the viscosity of food products, therefore offering mouthfeel to foods. This functionality linked the binding capacity of molecules especially with water; if they are substituted by the chemically sweetening additive like saccharin or aspartame, it does not affect the constancy of the food (James, 2009; Ahmad et al., 2012). One more substance will normally be added to provide the expected body or the mouthfeel to the food, for instance, hydrocolloids to the wanted consistency without adding the natural sugar (Belitz et al., 2009; Fennema, 1996). Table 2.6 exhibits the applications of polysaccharides in the food industry.

The defined functions of polysaccharides are dependent on their highly flexible properties. They might change from insoluble to those with the good swelling power and good solubility at diverse temperatures of solvent. The solutions might have very low viscosities at very high concentrations or might have extremely high viscosities at low concentrations. Some polysaccharides, at the low concentration, set into the thermoreversible gel. Whereas most gels melt at the elevated temperatures, some of the cellulose derivatives set into the gel (De Lumen, 1990; Davis, 1995).

2.6. ENZYMES

2.6.1. Classification

The systematic organization and description of enzymes centered on the specificity of reaction are based on the Nomenclature Committee of the IUB-MB (International Union of Biochemistry and Molecular Biology)–adopted instructions last edited in, 1992. All of the enzymes can be categorized into six major classes (Cravatt et al., 2008; Belitz et al., 2009):

a) Oxido reductases: catalyzing the oxidoreduction reactions;

b) Hydrolases: catalyzing hydrolytic cleavage of the C-O, C-O, C-N, and some of the other bonds;

c) Transferases: transporting a group from one compound to the other;

d) Lyases: clearing C-C, C-N, C-O, and some other bonds by removal;

e) Ligases or synthetases: catalyzing the linking together of two molecules coupled with the hydrolysis of the pyrophosphate bond in ATP or the similar triphosphate; and

f) Isomerases: catalyzing structural or geometric changes within a molecule.

Some other enzymes deprived of the suffix-as terminology are rennin, trypsin, and ficin.

Each enzyme possesses a systematic name and the number derived from the six classes mentioned above. Each class can normally be subdivided into the subclasses which more precisely denote the kind of reaction, for example, by identifying the electron donor of the oxidation-reduction reaction or by identifying the functional group supported over by transferase or cleaved by the hydrolase enzyme. Furthermore, each subclass is divided into sub-sub classes. For instance, sub-sub classes of the oxidoreductases are signified by identifying the acceptor which receives the electron from its corresponding donor (Haldane, 1957; Rejasse et al., 2007). An instance is enzyme ascorbic acid oxidase having systematic name L-ascorbate: oxygen oxidoreductase and the systematic number E. C. 1.1.10.3.3.

E.C.	1.	10.	3.	3.
Enzyme commission	Oxidoreductase	Subclass of oxidoreductase	Sub-subclass	The serial number of the enzyme inside the sub-subclass

2.6.2. Catalytic Actions

The catalytic activity of enzymes is dependent on their properties like enzyme specificity and kinetics. Enzyme kinetics is disturbed by enzyme inhibition and enzyme concentration. The enzyme's concentration is usually an enzyme-catalyzed reaction and is directly proportional to the reaction rate (Cornish-Bowden, 1995; Speers and Cravatt, 2004).

Michaelis and Menten were the first ones to present a mathematical analysis of the velocity of the enzyme-catalyzed reactions as disturbed by the concentration of the substrate (Figure 2.9). It is supposed that a midway enzyme-substrate complex is created, which is the vital feature of this theory. Furthermore, the supposition is made that the rate of conversion of the enzyme-substrate complex to the products of the reaction and the enzyme defines the rate of conversion of the substrate to reaction products. This equation is given as follows:

$$v = \frac{V[S]}{K_m + [S]} \quad (2)$$

Where, v →initial velocity; V → maximal velocity; [S] → total concentration of substrate; K_m → Michaelis-Menten constant.

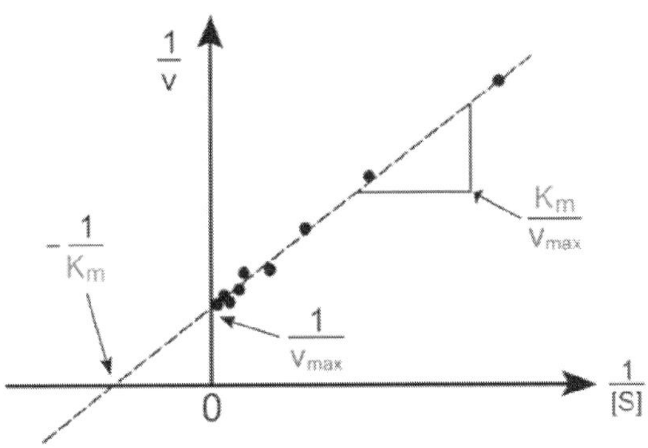

Figure 2.9: Linear plots of Michaelis-Menten equation

Source: http://commons.wikimedia.org/wiki/User:Diberri.

$$\frac{1}{v} = \frac{K_m}{V[S]} + \frac{1}{V} \qquad (3)$$

If 1/v is plotted v. s 1[S], then K_m/V is the slope and the intercept on the 1/v axis is 1/V. A straight line is obtained instead of the sigmoid curve as in the Michaelis-Menten. It is probable to calculate K_m (Figure 2.9).

The Michaelis-Menten equation can normally be analyzed by the impacts of the enzyme inhibitor on reaction kinetics. There are three classes of the enzyme inhibition: noncompetitive, competitive, and uncompetitive. All of the classes are reversible. In the competitive inhibition, an additional substance linked to substrate contests with the substrate to bind at the active site of the enzymes. This can prevent the enzyme activity in the creation of a complex with the normal substrates (Liceaga-Gesualdo and Li-Chan, 1999; Gbogouri et al., 2004). In the non-competitive inhibition, inhibitors don't bind at the enzyme's active site but at an additional site that often changes the enzyme as an outcome. In the uncompetitive inhibition, an inactive enzyme substrate inhibitor complex is made by the amalgamation with the enzyme-substrate complex, which is normally formed. This inhibitor itself doesn't react with the enzyme to create a complex. In the irreversible inhibition, some of the agents are capable to modify covalently and permanently a functional group of definite enzymes that is essential for catalysis, consequential in the irreversible inactivation of enzyme (Copeland, 2000; Cornish-Bowden, 1995).

2.6.3. Rate of Reaction

The enzymatic reaction rate is stated as the number of molecules of the substrate decomposed by the molecule of a certain enzyme per minute. As an instance, catalase decomposes 5,000 thousand molecules of the H_2O_2 in 60 seconds (Chobert et al., 1996; Kong et al., 2007). If the enzyme's MW isn't known or if the employed material isn't pure, it is accepted internationally that one unit of enzyme material utilized will generally catalyze the reaction one mmol of a substrate in one minute under governed conditions. Frank A. Lee (1983) defines the following factors that significantly affect the enzyme reaction rate:

a) Temperature;
b) Regeneration;
c) pH; and

d) Activators.

2.6.4. Uses in Foods

In vegetable processing, the key aim of the enzyme inactivation is to avert the quality worsening of vegetables. Steaming or blanching vegetables before the low-temperature storage can avoid their deterioration. Lee (1958) defines the freedom of fatty acids as an outcome of peroxidases discovered in the un-blanched peas during the storage stage at 18°C, while the compounds weren't made in the blanched peas. Conversely, blanching vegetables before the process of drying can remove the growth of haylike aroma and flavor (Wang et al., 1999; Kristinsson and Rasco, 2000). Whereas low water content safeguards food from spoilage by the microbial action, it doesn't stop deterioration as an outcome of enzyme activity. The increase in moisture content can result in an upsurge in enzyme activity (Northrop, 1932; Kim et al., 1990).

In the beverage industry, pectic enzymes are utilized to upsurge the production of juice and to completely extract the pigments from fruits. In alcoholic beverages, the microorganisms yield some enzymes that transform sugars into alcohol (Whitehurst and Law, 2002).

Enzymes are often isolated from parts of microorganisms and organisms. In the enzyme isolation, the richest accessible source should be utilized to lessen the amount of work mandatory to make the enzyme. The extraction must not be performed in extreme temperature, ionic strength, and pH, which triggers enzyme inactivity and disability. The isolation can normally be executed by chromatography with the help of some media like cross-linked polyacrylamide, ion-exchange, and Sephadex, or another current technique comprises the utilization of isoelectric focusing (Nagodawithana and Reed, 1993; Panyam and Kilara, 1996).

Enzymes can be beneficial due to the following characteristics:

(a) They are effective catalysts;
(b) The reaction can usually be stopped by applying the heat adequately to abolish the enzymes;
(c) Temperature, time, and pH can be utilized to govern the reactions;
(d) The activity of an enzyme can be standardized; and
(e) They are normally nontoxic and can generally be left in a product until it is essential to destroy them (Whitehurst and Law, 2002; Sathivel et al., 2003, 2005).

REFERENCES

1. Aasen, I. M., Markussen, S., Møretrø, T., Katla, T., Axelsson, L., and Naterstad, K., (2003). Interactions of the bacteriocins sakacin P and nisin with food constituents. *International Journal of Food Microbiology*, *87*(1–2), 35–43.
2. Ahmad, A., Anjum, F. M., Zahoor, T., Nawaz, H., and Dilshad, S. M. R., (2012). Beta-glucan: A valuable functional ingredient in foods. *Critical Reviews in Food Science and Nutrition*, *52*(3), 201–212.
3. Akoh, C. C., and Min, D. B., (2002). *Food lipids, Chemistry, Nutrition, and Biotechnology* (Vol. 1, pp. 1–50). Marcel Dekker.
4. Al-Muhtaseb, A. H., McMinn, W. A. M., and Magee, T. R. A., (2002). Moisture sorption isotherm characteristics of food products: A review. *Food and Bioproducts Processing*, *80*(2), 118–128.
5. Al-Muhtaseb, A. H., McMinn, W. A. M., and Magee, T. R. A., (2004). Water sorption isotherms of starch powders. Part 2: Thermodynamic characteristics. *Journal of Food Engineering*, *62*(2), 135–142.
6. Angibaud, P., (1986). Glass-forming tendency and stability of the amorphous state in the aqueous solutions of linear polyalcohols with four carbons. I. Binary systems water-polyalcohol. *Cryobiology, 23*, 453–469.
7. Ashbaugh, H. S., Truskett, T. M., Debenedetti, P. G., (2002). A simple molecular thermodynamic theory of hydrophobic hydration. *J. Chem. Phys., 116*, 2907–2921.
8. Barsotti, L., Dumay, E., Mu, T. H., Diaz, M. D. F., and Cheftel, J. C., (2001). Effects of high voltage electric pulses on protein-based food constituents and structures. *Trends in Food Science and Technology*, *12*(3–4), 136–144.
9. Belitz, H. D., Grosch, W., Schieberle, P., (2009). *Food Chemistry* (Vol. 1, pp. 1–50). 4th revised and extended edn. Springer, Berlin/Heidelberg.
10. Bhosale, R., and Singhal, R., (2007). Effect of octenylsuccinylation on physicochemical and functional properties of waxy maize and amaranth starches. *Carbohydrate Polymers*, *68*(3), 447–456.
11. Biliaderis, C. G., (1991). The structure and interactions of starch with food constituents. *Canadian Journal of Physiology and Pharmacology*, *69*(1), 60–78.
12. eBirch, G. G., and Parker, K. J. (1982). Intermediate moisture foods. *Applied Science, 1*, 1–30. London.

13. Boutron, P., Mehl, P., Kauffmann, A., Brady, G. W., and Romanov, W. J., (1960). Structure of water. *J. Chem. Phys., 32*, 106.
14. Boutron, P., Mehl, P., Kaufmann, A., and Angibaud, P., (1986). Glass-forming tendency and stability of the amorphous state in the aqueous solutions of linear polyalcohols with four carbons: I. Binary systems water-polyalcohol. *Cryobiology, 23*(5), 453–469.
15. Brady, G. W., and Romanow, W. J., (1960). Structure of water. *The Journal of Chemical Physics, 32*(1), 306–306.
16. Buettner, G. R., (1993). The pecking order of free radicals and antioxidants: Lipid peroxidation, a-tocopherol, and ascorbate. *Arch Biochem. Biophys., 300*, 535–543.
17. Cadden, A. M., (1988). Moisture sorption characteristics of several food fibers. *Journal of Food Science*, *53*(4), 1150–1155.
18. Capon, B., (1969). Mechanism in carbohydrate chemistry. *Chemical Reviews*, *69*(4), 407–498.
19. Cecil, R., and McPhee, J. R., (1959). The sulfur chemistry of proteins. In *Advances in Protein Chemistry* (Vol. 14, pp. 255–389). Academic Press.
20. Chalker, J. M., Bernardes, G. J., Lin, Y. A., and Davis, B. G., (2009). Chemical modification of proteins at cysteine: Opportunities in chemistry and biology. *Chemistry-An Asian Journal*, *4*(5), 630–640.
21. Chinachoti, P., (1995). Carbohydrates: Functionality in foods. *The American Journal of Clinical Nutrition*, *61*(4), 922S–929S.
22. Cho, S. S., Prosky, L., and Dreher, M., (1999). *Complex Carbohydrates in Foods* (Vol. 1, pp. 1–30). Marcel Dekker, New York.
23. Chobert, J. M., Briand, L., Guéguen, J., Popineau, Y., Larré, C., and Haertlé, T., (1996). Recent advances in enzymatic modifications of food proteins for improving their functional properties. *Food/Nahrung*, *40*(4), 177–182.
24. Chou, D. H., and Morr, C. V., (1979). Protein-water interactions and functional properties. *Journal of the American Oil Chemists' Society*, *56*(1), A53–A62.
25. Christakos, S., Gabrielides, C., and Rhoten, W. B., (1989). Vitamin D-dependent calcium binding proteins: chemistry, distribution, functional considerations, and molecular biology. *Endocrine Reviews*, *10*(1), 3–26.
26. Clements, J. A., (1971). Comparative lipid chemistry of lungs. *Archives*

of Internal Medicine, *127*(3), 387–389.

27. Copeland, R. A., (2000). Enzymes. *A Practical Introduction to Structure, Function, Mechanism, and Data Analysis* (Vol. 2, pp. 1–50). Wiley, New York.
28. Cornish-Bowden, A., (1995). *Fundamentals of Enzyme Kinetics* (Vol. 1, pp. 1–40). Portland Press, London.
29. Cravatt, B. F., Wright, A. T., and Kozarich, J. W., (2008). Activity-based protein profiling: From enzyme chemistry to proteomic chemistry. *Annu. Rev. Biochem.*, *77*, 383–414.
30. Damodaran, S., (2008). Amino acids, peptides, and proteins. In: Damodaran, S., Kirk, P., and Fennema, O. R., (eds.), *Fennema's Food Chemistry,* 4th edn., *1*, 219–329. CRC Press/Taylor and Francis Group, Boca Raton.
31. Damodaran, S., Parkin, K., and Fennema, O. R., (2008). *Fennema's Food Chemistry*. CRC Press, Boca Raton., *1*, 1-30.
32. Davis, E. A., (1995). Functionality of sugars: Physicochemical interactions in foods. *The American Journal of Clinical Nutrition*, *62*(1), 170S–177S.
33. De Lumen, B. O., (1990). Molecular approaches to improving the nutritional and functional properties of plant seeds as food sources: Developments and comments. *Journal of Agricultural and Food Chemistry*, *38*(9), 1779–1788.
34. De Man, J. M., (1999). *Principle of Food Chemistry,* (Vol. 3, pp. 1–55). Springer, New York.
35. Decker, E. A., and McClements, D. J., (2001). Transition metal and hydro peroxide interactions: An important determinant in the oxidative stability of lipid dispersions. *Inform., 12*, 251–255.
36. Decker, E. A., Xiong, Y. L., Calvert, J. T., Crum, A. D., and Blanchard, S. P., (1993). Chemical, physical, and functional properties of oxidized turkey white muscle myofibrillar proteins. *Journal of Agricultural and Food Chemistry*, *41*(2), 186–189.
37. Duckworth, R. B., Allison, J. Y., and Clapperton, H. A., (1976). The aqueous environment for change. In: Davies, R., *Birch, GG, and Parker, KJ*, 1, 80-89.
38. El Seoud, O. A., Koschella, A., Fidale, L. C., Dorn, S., and Heinze, T., (2007). Applications of ionic liquids in carbohydrate chemistry: A window of opportunities. *Biomacromolecules*, *8*(9), 2629–2647.

39. Erickson, M. D., and Frey, N., (1994). Property-enhanced oils in food applications. *Food Technol., 48*(11), 63–68.
40. Escarpa, A., González, M. C., Morales, M. D., and Saura-Calixto, F., (1997). An approach to the influence of nutrients and other food constituents on resistant starch formation. *Food Chemistry*, *60*(4), 527–532.
41. Fennema, O. R., (1973). Water and ice. In: *Low Temperature Preservation of Foods and Living Matter,* Marcel Dekker, New York. *1*, 1–44.
42. Fennema, O. R., (1978). Enzyme kinetics at low temperature and reduced water activity. In *Dry biological systems*, Academic Press., *1*, 297-322.
43. Fennema, O. R., (1996). *Food Chemistry*, CRC Press, Boca Raton. *1*, 1–40.
44. Ferry, J. D., (1980). *Viscoelastic Properties of Polymers* (Vol. 3, pp. 1–30). Wiley, New York.
45. Fischer, E., (1894). Influence of configuration on the action of enzymes. *Ber. Deut. Chem. Ges., 27*, 2985–2993.
46. Barbosa-Canovas, G. V., and Welti-Chanes, J. (1995). Food preservation by moisture control: fundamentals and applications. In *2nd International Symposium on the Properties of Water, Universidad de las Americas (USA)*, *2*, 1–40. Technomic Press, Lancaster.
47. Gbogouri, G. A., Linder, M., Fanni, J., and Parmentier, M., (2004). Influence of hydrolysis degree on the functional properties of salmon byproducts hydrolysates. *Journal of Food Science*, *69*(8), C615–C622.
48. Ghotra, B. S., Dyal, S. D., and Narine, S. S., (2002). Lipid shortenings: A review. *Food Res Int., 35*, 1015–1048.
49. Giese, J., (1996). Fats, oils, and fat replacers. *Food Technol., 50*(4), 78–84.
50. Goldfine, H. O. W. A. R. D., (1968). Lipid chemistry and metabolism. *Annual Review of Biochemistry*, *37*(1), 303–330.
51. Gontard, N., Duchez, C., Cuq, J. L., and Guilbert, S., (1994). Edible composite films of wheat gluten and lipids: water vapor permeability and other physical properties. *International Journal of Food Science and Technology*, *29*(1), 39–50.
52. Gulcin, I., (2012). Antioxidant activity of food constituents: An overview. *Archives of Toxicology*, *86*(3), 345–391.

53. Gutfreund, H., (1972). *Enzymes: Physical Principles* (Vol. 1, pp. 1–50). Wiley, London.

54. Haldane, J. B. S., (1957). Graphical methods in enzyme chemistry. *Nature*, *179*(4564), 832–832.

55. Hermansson, A. M., (1975). Functional properties of proteins for foods-flow properties. *Journal of Texture Studies*, *5*(4), 425–439.

56. Hermansson, A. M., (1977). Functional properties of proteins for foods-water vapor sorption. *International Journal of Food Science and Technology*, *12*(2), 177–187.

57. Hodge, J. E., (1953). Chemistry of browning reactions in model systems. *J. Agric. Food Chem.*, *1*, 928–943.

58. Hwang, K. S., Jang, S. P., and Choi, S. U., (2009). Flow and convective heat transfer characteristics of water-based Al_2O_3 nanofluids in fully developed laminar flow regime. *International Journal of Heat and Mass Transfer*, *52*(1–2), 193–199.

59. Hyung, H., and Kim, J. H., (2008). Natural organic matter (NOM) adsorption to multi-walled carbon nanotubes: Effect of NOM characteristics and water quality parameters. *Environmental Science and Technology*, *42*(12), 4416–4421.

60. Crowe. J. (2012). *Dry Biological Systems* (Vol. 1, pp. 297–322). Academic, New York.

61. Jambrak, A. R., Mason, T. J., Lelas, V., Herceg, Z., and Herceg, I. L., (2008). Effect of ultrasound treatment on solubility and foaming properties of whey protein suspensions. *Journal of Food Engineering*, *86*(2), 281–287.

62. James, L. E. A., (2009). Quinoa (Chenopodium quinoa Willd.): Composition, chemistry, nutritional, and functional properties. *Advances in Food and Nutrition Research*, *58*, 1–31.

63. Kaymak-Ertekin, F., and Sultanoğlu, M., (2001). Moisture sorption isotherm characteristics of peppers. *Journal of Food Engineering*, *47*(3), 225–231.

64. Kim, H. D., Kim, J. B., and Kim, M. H., (2006). Experimental study on CHF characteristics of water-ti02 nano-fluids. *Nuclear Engineering and Technology*, *38*(1), 61–68.

65. Kim, S. Y., Park, P. S., and Rhee, K. C., (1990). Functional properties of proteolytic enzyme modified soy protein isolate. *Journal of Agricultural and Food Chemistry*, *38*(3), 651–656.

66. Kinsella, J. E., and Melachouris, N., (1976). Functional properties of proteins in foods: A survey. *Critical Reviews in Food Science and Nutrition*, *7*(3), 219–280.
67. Kinsella, J. E., (1979). Functional properties of soy proteins. *Journal of the American Oil Chemists' Society*, *56*(3-Part 1), 242–258.
68. Kinsella, J. E., (1981). Functional properties of proteins: Possible relationships between structure and function in foams. *Food Chemistry*, *7*(4), 273–288.
69. Kinsella, J. E., (1982). Relationships between structure and functional properties of food proteins. *Food Proteins*, *1*, 51–103.
70. Kinsella, J. E., Rector, D. J., and Phillips, L. G., (1994). Physicochemical properties of proteins: texturization via gelation, glass, and film formation. In: Yada, R. Y., Jackman, R. L., and Smith, J. L., (eds), *Protein Structure-Function Relationships in Foods* (Vol. 1, pp. 1–21). Springer, New York.
71. Kong, X., Zhou, H., and Qian, H., (2007). Enzymatic preparation and functional properties of wheat gluten hydrolysates. *Food Chemistry*, *101*(2), 615–620.
72. Korycka-Dahl, M. B., Richardson, T., and Foote, C. S., (1978). Activated oxygen species and oxidation of food constituents. *Critical Reviews in Food Science and Nutrition*, *10*(3), 209–241.
73. Kristinsson, H. G., and Rasco, B. A., (2000). Fish protein hydrolysates: Production, biochemical, and functional properties. *Critical Reviews in Food Science and Nutrition*, *40*(1), 43–81.
74. Kwon, T. W., Menzel, D. B., and Olcott, H. S., (1965). Reactivity of malonaldehyde with food constituents. *Journal of Food Science*, *30*(5), 808–813.
75. Labuza, T. P., Cassil, S., and Sinskey, A. J., (1972). Stability of intermediate moisture foods. II. Microbiology. *J. Food Sci.*, *37*, 160–162.
76. Law, C. K., Lee, C. H., and Srinivasan, N., (1980). Combustion characteristics of water-in-oil emulsion droplets. *Combustion and Flame*, *37*, 125–143.
77. Lawler, P. J., and Dimick, P. S., (2002). Crystallization and polymorphism of fats. In: Akoh, C. C., and Min, D. B., (eds), *Food Lipids, Chemistry, Nutrition, and Biotechnology* (Vol. 1, pp. 1–22). Marcel Dekker, New York.

78. Lee, F. A., (1958). The effect of blanching on the carbonyl content of the crude lipid during the storage of frozen peas. *Food Res., 23*, 85–86.
79. Lee, F. A., (1983). *Basic Food Chemistry* (Vol. 1, pp. 1–45). The Avi Publishing, Westport.
80. Lewin, S., (1974). *Displacement of Water and its Control of Biochemical Reactions* (Vol. 1, pp. 1–40). Academic, London.
81. Liceaga-Gesualdo, A. M., and Li-Chan, E. C. Y., (1999). Functional properties of fish protein hydrolysate from herring (Clupea harengus). *Journal of Food Science*, *64*(6), 1000–1004.
82. Lieberman, H. R., Corkin, S., Spring, B. J., Growdon, J. H., and Wurtman, R. J., (1982). Mood, performance, and pain sensitivity: Changes induced by food constituents. *Journal of Psychiatric Research*, *17*(2), 135–145.
83. Linhardt, R. J., (2003). Claude S. Hudson Award address in carbohydrate chemistry. Heparin: structure and activity. *Journal of Medicinal Chemistry*, *46*(13), 2551–2564.
84. Lonchampt, P., and Hartel, R. W., (2004). Fat bloom in chocolate and compound coatings. Eur. *J Lipid Sci. Technol., 106*, 241–274.
85. Luck, W. A. P., (1981). Structures of water in aqueous systems. In: *Water Activity: Influences on Food Quality* (Vol. 1, pp. 1–50). Academic, New York.
86. Mala, G. M., and Li, D., (1999). Flow characteristics of water in microtubes. *International Journal of Heat and Fluid Flow*, *20*(2), 142–148.
87. Mala, G. M., Li, D., Werner, C., Jacobasch, H. J., and Ning, Y. B., (1997). Flow characteristics of water through a micro channel between two parallel plates with electro kinetic effects. *International Journal of Heat and Fluid Flow*, *18*(5), 489–496.
88. Marangoni, A., and Narine, S., (2002). *Physical Properties of Lipids* (Vol. 1, pp. 1–37). Marcel Dekker, New York.
89. Martins, S. I. F S., Jongen, W. M. F., and Boekel, M. A. J. S., (2001). A review of Maillard reaction in food and implications to kinetic modeling. *Trends Food Sci. Technol., 11*, 364–373.
90. Masters, P. M., and Friedman, M. E. N. D. E. L., (1980). Amino acid racemization in alkali-treated food proteins-chemistry, toxicology, and nutritional consequences. In *ACS Symposium Series* (Vol. 123, pp. 165–194).

91. Mazzio, E. A., Harris, N., and Soliman, K. F., (1998). Food constituents attenuate monoamine oxidase activity and peroxide levels in C6 astrocyte cells. *Planta Medica, 64*(07), 603–606.
92. McClements, D. J., (2002). Theoretical prediction of emulsion color. *Adv. Colloid Interface Sci., 97*, 63–89.
93. McClements, D. J., (2004). *Food Emulsions: Principles, Practice, and Techniques* (Vol. 2, pp. 1–40), CRC Press, Boca Raton.
94. McGrady, J., (1994). More-healthful fats and oils. *Food Technol., 48*(11), 148.
95. Michaelis, L., and Menten, M. L., (1913). The kinetics of invertase action. *Biochemistry, 49*, 333–369.
96. Min, D. B., and Boff, J. M., (2002). Lipid oxidation in edible oil. In: Akoh, C. C., and Min, D. B., (eds.), *Food Lipids, Chemistry, Nutrition, and Biotechnology* (Vol. 1, pp. 1–36). Marcel Dekker, New York.
97. Moran, D. P. J., (1994). Fats in spreadable products. In: *Fats in food products* (Vol. 1, pp. 1–45). Blackie Academic and Professional, London.
98. Morris, V. J., (1990). Biotechnically produced carbohydrates with functional properties for use in food systems. *Food Biotechnology, 4*(1), 45–57.
99. Murakami, H., Watanabe, Y., and Nakashima, N., (1996). Fullerene lipid chemistry: Self-organized multibilayer films of a C60-bearing lipid with main and sub-phase transitions. *Journal of the American Chemical Society, 118*(18), 4484–4485.
100. Murphy, O., (2001). Non-polyol low-digestible carbohydrates: Food applications and functional benefits. *British Journal of Nutrition, 85*(S1), S47–S53.
101. Nagodawithana, T., and Reed, G. E., (1993). Enzymes in Food Processing, (Vol. 3, pp. 1–30). Academic, New York.
102. Nelson, D. L., and Cox, M. M., (2013). Lehninger Principles of Biochemistry (Vol. 1, pp. 1–44). Worth Publishers, New York.
103. Nielsen, S. S., (2003). *Food Analysis,* (Vol. 1, pp. 1–40). New York.
104. Northrop, J. H., (1932). Crystalline trypsin IV. reversibility of the inactivation and denaturation of trypsin by heat. *The Journal of General Physiology, 16*(2), 323–337.
105. O'Keefe, S. F., (2002). Nomenclature and classification of lipids. In:

Akoh, C. C., and Min, D. B., (eds.), *Food Lipids, Chemistry, Nutrition, and Biotechnology* (Vol. 1, pp. 1–34). Marcel Dekker, New York.

106. Panyam, D., and Kilara, A., (1996). Enhancing the functionality of food proteins by enzymatic modification. *Trends in Food Science and Technology*, *7*(4), 120–125.
107. Patino, J. M. R., Sanchez, C. C., and Niño, M. R. R., (2008). Implications of interfacial characteristics of food foaming agents in foam formulations. *Advances in Colloid and Interface Science*, *140*(2), 95–113.
108. Peng, X. F., Peterson, G. P., and Wang, B. X., (1994a). Frictional flow characteristics of water flowing through rectangular microchannels. *Experimental Heat Transfer An International Journal*, *7*(4), 249–264.
109. Peng, X. F., Peterson, G. P., and Wang, B. X., (1994b). Heat transfer characteristics of water flowing through microchannels. *Experimental Heat Transfer An International Journal*, *7*(4), 265–283.
110. Pérez-Jiménez, J., and Saura-Calixto, F., (2006). Effect of solvent and certain food constituents on different antioxidant capacity assays. *Food Research International*, *39*(7), 791–800.
111. Pleiss, J., Fischer, M., and Schmid, R. D., (1998). Anatomy of lipase binding sites: The scissile fatty acid-binding site. *Chem. Phys. Lipids*, *93*, 67–80.
112. Pszczola, D. E., (2000). Putting fat back into foods. *Food Technol.*, *54*(12), 58–60.
113. Rejasse, B., Lamare, S., Legoy, M. D., and Besson, T., (2007). Influence of microwave irradiation on enzymatic properties: Applications in enzyme chemistry. *Journal of Enzyme Inhibition and Medicinal Chemistry*, *22*(5), 519–527.
114. Robinson, N., and Eglinton, G., (1990). Lipid chemistry of Icelandic hot spring microbial mats. *Organic Geochemistry*, *15*(3), 291–298.
115. Rousseau, D., and Marangoni, A. G., (2002). Chemical interesterification of food lipids: Theory and practice. In: Akoh, C. C., and Min, D. B., (eds.), *Food Lipids, Chemistry, Nutrition, and Biotechnology* (Vol. 1, pp. 1–26). Marcel Dekker, New York.
116. Sathivel, S., Bechtel, P. J., Babbitt, J., Smiley, S., Crapo, C., Reppond, K. D., and Prinyawiwatkul, W., (2003). Biochemical and functional properties of herring (Clupea harengus) byproduct hydrolysates. *Journal of Food Science*, *68*(7), 2196–2200.

117. Sathivel, S., Smiley, S., Prinyawiwatkul, W., and Bechtel, P. J., (2005). Functional and nutritional properties of red salmon (Oncorhynchus nerka) enzymatic hydrolysates. *Journal of Food Science*, *70*(6), c401–c406.
118. Schippers, J. C., and Verdouw, J., (1980). The modified fouling index, a method of determining the fouling characteristics of water. *Desalination*, *32*, 137–148.
119. Schmidt, S. J., (2004). Water and solids mobility in foods. *Adv. Food Nutr. Res.*, *48*, 1–101.
120. Scott, W. J., (1957). Water relations of food spoilage organisms. *Adv. Food Res.*, *7*, 83–127.
121. Serdarevich, B., (1967). Glyceride isomerizations in lipid chemistry. *Journal of the American Oil Chemists' Society*, *44*(7), 381–393.
122. Seu, K. J., Cambrea, L. R., Everly, R. M., and Hovis, J. S., (2006). Influence of lipid chemistry on membrane fluidity: Tail and head group interactions. *Biophysical Journal*, *91*(10), 3727–3735.
123. Shao, X., and Li, Y., (2013). Application of low-field NMR to analyze water characteristics and predict unfrozen water in blanched sweet corn. *Food and Bioprocess Technology*, *6*(6), 1593–1599.
124. Singh, H., (1991). Modification of food proteins by covalent cross linking. *Trends in Food Science and Technology*, *2*, 196–200.
125. Slade, L., and Levine, H., (1995). Polymer Science Approach to Water Relationships in Foods. *Food preservation by moisture control*, 1, 33-132 (Vol. 1, pp. 1–50).
126. Soottitantawat, A., Yoshii, H., Furuta, T., Ohgawara, M., Forssell, P., Partanen, R. and Linko, P., (2004). Effect of water activity on the release characteristics and oxidative stability of D-limonene encapsulated by spray drying. *Journal of Agricultural and Food Chemistry*, *52*(5), 1269–1276.
127. Speers, A. E., and Cravatt, B. F., (2004). Profiling enzyme activities in vivo using click chemistry methods. *Chemistry and Biology*, *11*(4), 535–546.
128. Steinke, M. E., and Kandlikar, S. G., (2004). An experimental investigation of flow boiling characteristics of water in parallel microchannels. *J. Heat Transfer*, *126*(4), 518–526.
129. Stephen, A. M., (1995). *Food Polysaccharides and Their Applications* (Vol. 1, pp. 1–50). Marcel Dekker, New York.

130. Stick, R. V., and Williams, S. J., (2009). *Carbohydrates: The Essential Molecules of Life* (Vol. 1, pp. 1–40). Elsevier, Amsterdam.
131. Timms, R. E., (1991). Crystallization of fats. *Chem. Ind., 1*, 342–349.
132. Tvaroŝka, I., and Bleha, T., (1989). Anomeric and exo-anomeric effects in carbohydrate chemistry. In *Advances in Carbohydrate Chemistry and Biochemistry* (Vol. 47, pp. 45–123). Academic Press.
133. Uniacke-Lowe, T., Huppertz, T., and Fox, P. F., (2010). Equine milk proteins: Chemistry, structure, and nutritional significance. *International Dairy Journal*, *20*(9), 609–629.
134. Ustunol, Z., (2015). Amino acids, peptides, and proteins. In: Ustunol, Z., (ed.), A*pplied Food Protein Chemistry* (1st edn., Vol. 1, pp. 11–21) Wiley, Chichester.
135. Vaclavik, V. A., and Christian, E. W., (2014). *Essential of Food Science* (Vol. 4, pp. 1–30), Springer, New York.
136. Velázquez, M., and Feirtag, J. M., (1999). Helicobacter pylori: Characteristics, pathogenicity, detection methods and mode of transmission implicating foods and water. *International Journal of Food Microbiology*, *53*(2–3), 95–104.
137. Walstra, P., (2003). *Physical Chemistry of Foods* (Vol. 1, pp. 1–30). Marcel Dekker New York.
138. Wang, M., Hettiarachchy, N. S., Qi, M., Burks, W., and Siebenmorgen, T., (1999). Preparation and functional properties of rice bran protein isolate. *Journal of Agricultural and Food Chemistry*, *47*(2), 411–416.
139. Wang, N., and Brennan, J. G., (1991). Moisture sorption isotherm characteristics of potatoes at four temperatures. *Journal of Food Engineering*, *14*(4), 269–287.
140. Weiss, T. J., (1983). *Food Oils and Their Uses* (Vol. 2, pp. 1–30). AVI Publishing, Westport.
141. Whistler, R. L., and BeMiller, J. N., (1993). Industrial Gums (Vol. 3, pp. 1–30). Academic, San Diego.
142. Whistler, R. L., (1964). Methods in carbohydrate chemistry. *Methods in Carbohydrate Chemistry, 4*, 1–30.
143. White, G. W., and Cakebread, S. H., (1966). The glassy state in certain sugar-containing food products. *J Food Technol., 1*, 73–82.
144. White, G. W., and Cakebread, S. H., (1969). Importance of the glassy state in certain sugar-containing food products. *Food Sci. Technol.*

Proc. Int. Congr., 1, 227–235.

145. Whitehurst, R. J., and Law, B. A. E., (2002). *Enzymes in Food Technology* (2nd edn., Vol. 2, pp. 1–30) CRC Press, Boca Raton.
146. Wieser, H., (2007). Chemistry of gluten proteins. *Food Microbiology, 24*(2), 115–119.
147. Wiester, M. J., Ulmann, P. A., and Mirkin, C. A., (2011). Enzyme mimics based upon supramolecular coordination chemistry. *Angewandte Chemie International Edition, 50*(1), 114–137.
148. Wilding, P., Lillford, P. J., and Regenstein, J. M., (1984). Functional properties of proteins in foods. *Journal of Chemical Technology and Biotechnology. Biotechnology, 34*(3), 182–189.
149. Williams, M. L., Landel, R. F., and Ferry, J. D., (1955). The temperature dependence of relaxation mechanisms in amorphous polymers and other glass-forming liquids. *J. Am. Chem. Soc., 77*, 3701–3707.
150. Withers, S., (2015). *Anomeric Centre (alpha and beta)* (Vol. 1, pp. 1–30). CAZypedia. http://www.cazypedia.org/index.php/Anomeric_ centre_ (alpha_and_beta) (accessed on 3 March 2020).
151. Wnson, S. B., Williams, H. J., Frankie, G. W., and Shrum, G., (1997). Floral lipid chemistry of Byrsonima crassifolia (Malpigheaceae) and a use of floral lipids by Centris bees (Hymenoptera: Apidae). *Biotropica, 29*(1), 76–83.
152. Xiao, F., Simcik, M. F., and Gulliver, J. S., (2012). Partitioning characteristics of perfluorooctane sulfonate between water and foods. *Archives of Environmental Contamination and Toxicology, 62*(1), 42–48.
153. Zayas, J. F., (1997). *Functionality of Proteins in Food* (Vol. 1, pp. 1–30). Springer, Berlin/Heidelberg/New York.
154. Zhou, Z., Robards, K., Helliwell, S., and Blanchard, C., (2002). Composition and functional properties of rice. *International Journal of Food Science and Technology, 37*(8), 849–868.
155. Zhuang, H., Barth, M. M., and Hildebrand, D., (2002). Fatty acid oxidation in plant lipids. In: Akoh, C. C., and Min, D. B., (eds.), *Food Lipids, Chemistry, Nutrition, and Biotechnology* (Vol. 1, pp. 1–30). Marcel Dekker, New York.
156. Zilversmit, D. B., (1968). The surface coat of chylomicrons: Lipid chemistry. *Journal of Lipid Research, 9*(2), 180–186.

Chapter 3

Sampling and Sample Preparation for Food Analysis

CONTENTS

3.1. INTRODUCTION

The concept of food is probably the first problem appeared while dealing with food science. Specialized literature came up with several possible definitions for this concept. Some definitions emphasize on its constitution (typically, water, fats, protein, and carbohydrates), others focus on human behavior to get food items (which has nowadays changed to fishing, ranching, and farming in many civilizations from formerly used methods such as gathering and hunting). In other scenarios, definitions emphasize the essence of the problem itself and the presumed benefices related to its utilization. Moreover, one should identify that stated definitions of food concept highly depend on culture as well. Plants, minerals, water, animals, or other categories like fungus, processed, elaborated, and fermented products may be considered as sources of food. By keeping into mind some of above viewpoints, the definition of food could be stated as any product or substance, solid or liquid, processed or elaborated, or natural that, is consumed or drunk by human beings as nutrient and amusement because of its features, uses, constituents, preservation state, and preparation.

Regardless of any definition adopted, everyone agrees that food is a complicated combination of a large number of chemical compounds. Moreover, it is a consensus that food key features are its physical properties and chemical combination. Because of these key features of food, one can find out its nutritional value and hygienic state. Its popularity in consumers and users can also be estimated by using these features. That is why these two features are very important for food analysts and legislators.

There are several varying reasons for analyzation of foodstuffs. Different government institutes make sure that foodstuff is legalized, perfectly labeled and validated. Economic losses or consumers health damage can be controlled by before time detection of illegal practices and adulteration. The analysis of food is also of utmost significance for the food industry which evaluates the standard of the original material and maintenance during the processing of food, its preservation, and transportation procedures. Many food scientists are working hard on improving different methodologies that are used to manage all the discussed aspects. Also, modern analytical processes are in the development to lower the Maximum Residue Levels (MRLs) of poisonous chemicals, for detailed categorization of food products, add new chemicals in legislation and make such new foodstuffs which will be added value. Finally, in modern times, Quality of food is a growing concern for consumers. The reason behind this concern is news of contamination of food by adding different forbidden and toxicants. Also, health and diet accepted relationship and growing demand in consumer for nutritional food products is making these concerns more significant. The demand for nutritional food products is justification for the development and incorporation of new food elements. The short-term and long-term effects of these added food elements on foodstuffs should be tested.

It is clear from thoughts that food analysis is a very extensive field that is in continuous evolution. It includes analysis and chemical determination of different characters and with widely varying objectives. Such differences are the reason for different methods in operation for food analysis. As exhibited in Figure 3.1, the scope of these methods varies from subjective (for example, organoleptic determinations) to objective methods based on microbiologic, microscopic, chemical, and physical determinations. Other methods, e.g., personal questionnaires and biological determinations are also utilized. This book evaluates the modern and latest developments in chemical approaches and will give attention to the usage of state-of-the-art instrumental analytical techniques that have been recently adopted in this expanding/growing research field.

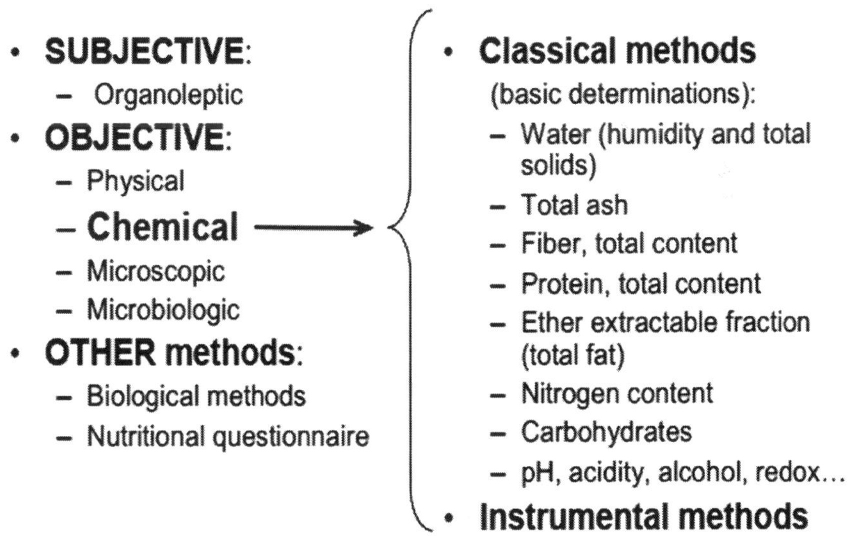

Figure 3.1: Methods for food analysis.

Source: https://www.sciencedirect.com/book/9780123848628/chemical-analysis-of-food-techniques-and-applications.

3.2. TYPES OF SAMPLES AND THE ANALYTICAL PROCEDURE

Food analysis requires chemical analysis at different stages and for different objectives. As discussed before, for traditional foods, different chemical methods and determination are implemented on raw materials and independent ingredients transform them into processed foods and end-products. Intermediate products are added to ensure the quality of food. Such analysis is particularly pertinent in the implementation and developments of new methods and preservation procedures or during the development of new chemical formulas and food products. Other analytical methods, chemical determination of food products consist of several steps. These steps have a huge impact on the authenticity of the generated data (Figure.3.2).

On-site determination is also a possibility in some scenarios, samples are transported to the research laboratory and chemical analysis is accomplished. Every occasion, it should be taken care that how many samples are acquired and what should be their size. Along with that location of samples matters to guarantee the representativeness of all subsamples. Whether purposeful or aimless, notable consideration should be made during the sampling protocol for acquiring relevant and understandable results from the analytical process. Sampling is a complicated method. Initially, sampling relied on the form of the sampled matrix (liquid or solid), its dimension (as a part or as a complete entity) and objective of the analysis (e.g., trace analysis or analysis of foremost parts). In few scenarios, the minimum number of samples necessary and method to start the specific analysis is undoubtedly mentioned in following regulations [In animals and plants origin products, for analysis of pesticide leftovers look at, e.g., [(2002/63/EC, 2002) and (90/642/EEC, 1993)]. In other situations, if the representativeness of the sampling method is guaranteed then methods described in regulatory text or substitute methods can be embraced. Detail discussion on this complicated matter is beyond the scope of this chapter. Samples should be unchanged while transferring and followed by preservation and analysis.

Before instrumental analysis for target component initiate, samples undergo several handlings and processes in the Lab. These processes are generally known as sample preparation. The essence and number of processes and operations are determined by these factors: target component type and its foreseen concentration value, potential matrix interference constituents, responsiveness, and selectiveness of analytical procedure which is chosen for ultimate detection and selection. Samples are undergone preparation which includes different tasks from tagging and mechanical handling and uniformity to the measurement of samples either by weight or by volume. Moreover, all operations performed to divide the matrix construction to conduct enrichment, segregation, and fractionation of the target component are considered as part of sample preparation. The process to check tested analyte is suitable for detector and process to increase the reactivity of the detector are also performed during sample preparation.

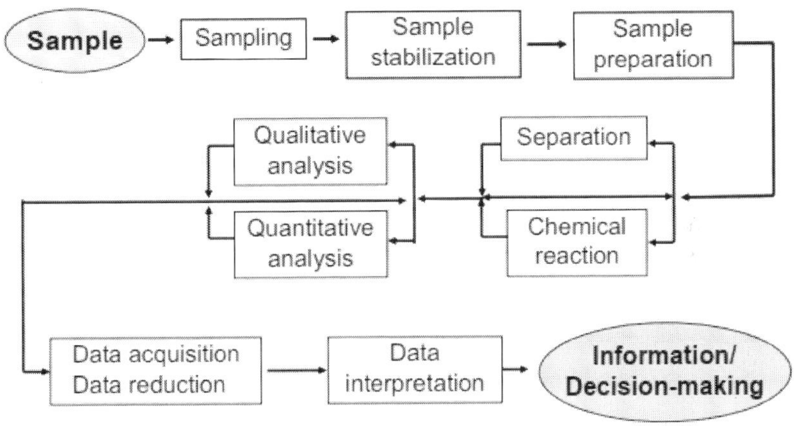

Figure 3.2: Overview of an analytical process.

Source: https://www.sciencedirect.com/book/9780123848628/chemical-analysis-of-food-techniques-and-applications.

Food Ingredients and food adulterate which are essential for chemical analysis; complete overview of these foodstuffs is covered in the following list.

1. Organometallic and metallic species;
2. Process and storage residues;
3. Migrants from packaging materials;
4. Phyto-toxins and mycotoxins;
5. Contaminants PBDEs, PAHs, PCDD/Fs, PCBs, mineral oils, phthalates;

   Veterinarian drugs and pesticides;
6. Food contaminants and additives;
7. Polyphenols and antioxidants;

8. Vitamins, carotenoids, carbohydrates, proteins, and lipids;
9. Semi-volatiles or non-volatiles;
10. Off-flavors;
11. Fragrances and flavor; and
12. Volatile components.

In many times, target analytes are also related to routine controls. Target compounds add from volatile compounds to metals and organometallic. After that fragrances and flavor include. Also, off-flavors introduce, they can initiate issues with undesirable foodstuffs. New research, quality, and legal reasons motivate to analyze several minor and main constituents with added functional and nutritional value like vitamins, carbohydrates, proteins, lipids, and antioxidants. Also, a large kind of contaminants, residues, and food additives of different nature compel that, food items should be monitored regularly to exercise the present regulations and legislations. Research efforts are continuously carried out to upgrade the responsiveness and accuracy of the analytical procedure to detect a specific component. So that the social pressure of providing safe and secure food can be eliminated.

There are very few determination techniques that are directly used for determination (for example spectroscopy determination with chemometrics apply for the determination of food constituents. Before instrumental analysis, sample preparation desire for all analytes specified in Table 3.1, regardless of the procedure adopted for ultimate Separation-plus-detection. In the straightforward scenario, quantitative (i.e., nonselective, and exhaustive) derivation of the entangled compound of interest from a matrix is carried out. A final purified analyte is extracted after segregation from other coextracted components to make sure analyte accurate detection. Other than that application, many analytical steps are time-wasting and exhausting. So, these steps are implemented in off-line mode. Normally, a decline in the amount of target compound enhances complications of the operation and due to continuous handling of the extract, there are some chances of contamination and loss of the analyte. Recently, the struggle has been dedicated to overcoming these obstacles; consequently, more impressive, versatile, and rapid extraction methods have been developed. Other research fields like molecular and environmental analysis also contribute to these developments. Modern techniques develop for the volatile component analysis (Table 1.2) are Stir-Bar-Sportive extraction (SBSE), solid-phase microextraction (SPME) and Automated Purge-and-Trap (P and T), Nonvolatile, and semi-volatile analytes which are in liquid form are determined by adapting several solvent-based microextraction methods. Viscous and semi-solid samples are analyzed by using fluid/solvent extraction techniques, for example, ultrasound-assisted extraction (USE), microwave-assisted extraction (MAE) subcritical water extraction (SWE), pressurized liquid extraction (PLE) and supercritical fluid extraction (SFE). Other than that matrix solid-phase dispersion (MSPD) and microfluidic devices like molecular techniques, Real-time PCR and DNA arrays are also adopted techniques. In this chapter, only state-of-the-art and trendy analytical techniques for food analysis are discussed by using selected representative examples. Other techniques will be discussed in the upcoming chapters of this book.

3.3. THE TENDENCY IN SAMPLE PREPARATION

The reason for the implementation of Physico-chemical treatment is to separate the analytes from main matrix constituents so that, the instrumental determination could not be obstructed and to enhance the amount of the analyte during distillation. That second reason is regarded as a part of the sample preparation. One can reach the point that most official and traditional sample preparation procedures (AOAC, 1990; Nollet, 1996) is applied are lengthy, tedious, and contain multiple steps. These numerous steps cause contamination, degradation, and even loss of the target analytes. One can conclude that sample preparation is an important part of food analysis and it affects the analysis in many ways like (i) Authenticity of final out, (ii) Estimation of cost by measuring sorbents and solvent utilization, and (iii) Time consumption for the process.

Sample preparation creates hindrance in performing the analytical procedures on food. It is calculated that almost 60–80% of the execution expenditure and work activity in the analytical procedures are used in constructing samples for selecting instrument determination. Also, that part obliged for more than 50% of the error in results. These facts show that why continuous efforts were made in recent years to establish a greener, cost-effective, automated, and faster alternative to conventional protocols.

3.3.1. Solid-Phase Microextraction (SPME)

Most of these requirements are fulfilled by miniaturized Solid-Phase Microextraction (SPME).

Table 3.1: Overview of Analytical Techniques

The Base of the technique **Name of the technique (acronym)**
Enhanced Fluid/Solvent Extraction Ultrasound-assisted Extraction (USE)
Microwave-assisted extraction (MAE)
Subcritical Water Extraction (SWE)
Pressurized Liquid Extraction (PLE)
Supercritical Fluid Extraction (SFE)
Matrix Solid Phase Dispersion Matrix Solid-Phase Dispersion (MSPD)
Thermal Desorption
Stir Bar Sorptive Extraction (SBSE)
Solid Phase Microextraction (SPME)
Sorption Extraction
Restricted Access Medium (RAM)
Molecular Imprinted Solid Phase Extraction (MISPE)
Dispersive Solid Phase Extraction (DSPE)
Solid Phase Microextraction (SPME)
Single Short Column (SSC)
Fiber-in-tube Solid Phase Extraction (fiber-in-tube SPE)
In-tubeSolid Phase Microextraction (In-tube SPME)
Solid Phase DynamicExtraction
Open-tubular-coated Capillaries
Solid Phase Extraction (SPE)
Solvent extraction Dispersive Liquid-Liquid Microextraction (DLLME) Extracting Syringe (ESY)
Liquid Phase Microextraction (LPME)
Single Drop Microextraction (SDME)
In-vial Liquid-Liquid Extraction (in-vial LLE)
Liquid-Liquid Extraction (LLE)
The purge of volatile compounds Simultaneous Distillation-extraction (SDE)
Direct Thermal Desorption (DTD)
Programmed Thermal Vaporization (PTV)
Purge and Trap (PandT)
Static and Dynamic Headspace (S/D HS)

In solid-phase microextraction technique, the analyte is consumed by fused silica fiber which is coated by a sorbent suitable for the analyte. Fused silica fiber is immersed in the sample liquid or placed in the head-space (HS) of the sample. SPME has been abundantly implemented in several applications inclusive of food analysis since its inception by Pawliszyn's group (Arthur and Pawliszyn, 1990) in 1990. It is a solvent-free preconcentration technique. Initially, volatile analytes from solid, semi-solid, and liquid have been concentrated to liquid by implementing it. It is a comparatively fast, selective, and simple technique. There are many applications of SPME for example assessment of gapes origin (Rocha et al., 2007), oxidation of lipids and degeneration of proteins at the time of storing soup particles (Raitioa et al., 2011). In this first mentioned technique, the mono-terpenoid layer of *Vitis vinifera* L. cv. 'Fernao-Pires' white grape was established in HS mode by fusing SPME silica fiber which was glazed by Carbowax-divinylbenzene. In grapes, 56 monoterpenoids are detected by extensive two-dimensional gas chromatography (GC) with time-of-flight mass spectrometry (GC × GC-ToF MS) integrated with-SPME. From these 56 monoterpenoids, 20 were freshly determined. In Fig. 3.3, results of regular examples are shown. Author closing remarks are: Variety-related genes encode the synthesis of monoterpenoids which are secondary metabolites. Grape varietal nature traces by the terpenoids layer.

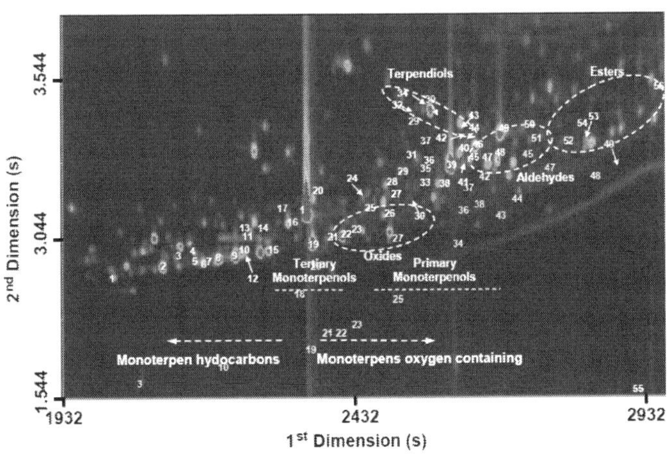

Figure 3.3: Clusters or bands created by structurally relevant compounds are indicated. Ions m/z 93, 121 and 136GC × GC plot.

Source: https://www.sciencedirect.com/book/9780123848628/chemical-analysis-of-food-techniques-and-applications.

3.3.2. Stir-Bar-Sorptive Extraction (SBSE)

The latest sample preparation techniques like SPME or headspace (HS) and traditional simultaneous distillation-extraction (SDE) are different as compare to stir-bar-sorptive extraction (SBSE). Because SBSE is beneficial in detecting unfamiliar traces (Ridgway et al., 2010). Also, SBSE utilizes a lesser amount of solvent which decreases the contamination risk from outward lab resources and enhances the detection in contrast to SPME and HS. One can say that the results presented by SBSE are more satisfactory. But, the determination of hexanal and methyl methacrylate is not suitable by this method. In Olariu et al. (2010), a comparison between traditional sample preparation methods, SPE, SPME, and SBSE is carried out in terms of their limitation and advantages.

3.3.3. Liquid-liquid Extraction

One of the most formal methods is liquid-liquid extraction (LLE). But, to enlarge the scope of applications by including new target compounds, some of these methods are revised repeatedly (Mol et al., 2007). The uncomplicatedness of the LLE procedure would give rise to their adoption of carrying some advanced solvent microextraction methods which are comparably a straightforward task by reducing the actual method. Practically, this technique can be tricky. But, its frequent execution and great sensitivity of advance instrumental methods in food analysis are the evident reasons behind the use of it [see, e.g., Asensio-Ramos et al. (2011). The first developed solvent-based microextraction technique was single drop microextraction (SDME). Still, it has been abundantly implemented in food analysis. Some of the common examples of SDME are listed below.

3.3.4. Single-Drop Microextraction

Table 3.2: Typical Applications Embracing SDME

Sample type	Analytes	Mode	Solvent (mL)	LOD[a] (mg/L, ng/g)	Time (Min)	Reference
SDME *Two-phase system*						

Chocolate, Filtrated beverages and Degassed	Caffeine	DDME[f]	Chloroform (0.5)	4.0	5	Shrivas and Wu (2007)
Digested Fat Free milk powder	Lead	DI	[BMIM][PF_6] [e](7)	0.015	7	Manzoori et al. (2009)
Mineralized rice flour	Cadmium	DI	Dithizone (0.01 M) in chloroform (3)	0.7×10^{-3}	10	Fanan-dZhou(2006)
Fruit extracts, Fruit juices	Organic acids	DI	Hexyl acetate (2.5)	0.6–160	20	Saraji and-Mousavinia (2006)
Goldfish extract	3 OCPs[d]	DI	Toluene (0.6)	0.5	7	Shrivas and Wu (2008)
Orange juice	7 OPPs[b]	DI[c]	Toluene (1.6)	0.98–2.2	15	Zhao et al. (2006)
Three-phase system						
Green tea extract	Adenine	LLLME[h]	EtOAc[g](200)	2	10	Zhu et al. (2010)
Mussel extract	3 Butyltin compounds	HS	n-C_{10} (2)	$(0.8–800) \times 10^{-3}$	5	Xiao et al. (2008)
Cuminum cymimum L extract	26 Essential oil compounds	HS	n-C_{17} (2)	$(6.7–15) \times 10^{-6}$	5	Wang et al. (2009)
Diluted wine	Anisols	HS	1-Octanol (2)	$(6.1–8.1) \times 10^{-3}$	25	Martendal et al. (2007)
Beer	Alcohols	HS	Ethylene glycol (1)	(3.8–52) 3 × 10	6	Tankeviciute et al. (2001)
HF(2/3)LPME[i]						
Buffered bovine milk	3 Tetracycline antibiotics	HF(3) LPME[j]	Aliquat 336 in 1-octanol H_3PO_4, 0.1 M	0.5–1.0 HF(3) LPME	35	Shariati et al. (2009)

Mineralized oyster (reference material)	Nickel, lead	[HMIM][PF_6][k](8)	HF(2)LPME	0.02–0.03	27	Abulhassani et al. (2010)
Diluted milk, beer, juice	Volatile organic selenium species	1-Decanol (3)	HF(2)LPME	(57–65) × 10^{-3}	5	Ghasemi et al. (2011)
Filtrated fruit juices	7 Phenolic acids	Hexyl acetate[l] NaOH, 0.02 M (8)	HF(3)LPME	0.01–2.0	30	Saraji and Mousavi (2010)
Aqueous green tea and tea leaves extract	6 Organosulfur pesticides	o-Xylene (5)	HF(2)LPME	0.21–3.0	35	Xiong and Hu (2008)
Filtrated orange juice	2 Fungicides	2-Octanone[m] HCl, 10 mM (20)	HF(3)LPME	0.05–0.10	30	Barahona et al. (2010)
Alcoholic beverages	51 Multiclasspesticides	1-Octanol (-)	HF(2)LPME	0.01–2.0	50	Plaza Bolanos et al., (2008)
DLLME						
Mineralized rice, tea, defatted milk powder	Cadmium, cobalt	–	Carbon tetrachloride (100)	(0.01–0.5) × 10^{-3}	3	Wen et al. (2011)
Extracted and purified porcine tissue	Clenbuterol	–	Tetra-chloromethylene (150)	0.07	5	Liu et al. (2011a)
Extracted and purified food extracts (milk, egg yolk, olive oil)	Cholesterol	–	Carbon tetrachloride (35)	0.01	1–2	Daneshfar et al. (2009)

Milk	3 Phenyl-urea pesticides, 4 triazines	–	[HMIM][PF_6] (60)	0.46–2.0	7	Gao et al. (2010)
Extracted and purified food extracts (milk, egg yolk, olive oil)	Cholesterol	–	Carbon tetrachloride (35)	0.01	1–2	Daneshfar et al. (2009)
Banana extract	8 Multiclass pesticides	–	[HMIM] [PF_6] (88)	0.32–4.7	20	Ravelo-Perez et al. (2009)
Centrifuged apple juice	24 Multiclass pesticides	Carbon tetrachloride (100)	Carbon tetrachloride (100)	0.06–2.2	3	Cunha et al. (2009)

[a]Limits of detection; [b]Organophosphorus pesticides; [c]Direct immersion; [d]Organochlorine pesticides; [e]1-Butyl-3-methylimidazolium hexafluorophosphate; [f]Drop-to-drop microextraction; [g]Ethyl acetate; [h]Liquid-liquid-liquid microextraction; [i]Hollow-fiber liquid-phase microextraction; [j]For this technique, the extraction solvent column corresponds to the acceptor phase (followed, when applicable, by the back-extraction phase). Otherwise specified, the donor phase is the corresponding buffered sample or sample extract; [k]1-Hexyl-3-methylimidazolium hexafluorophosphate; [l]Polychlorinated biphenyls; [m]Polybrominated diphenyl ethers.

SDME can be implemented in the latest liquid-liquid-liquid microextraction (LLLME) or HS mode as a three-phase system and drop-to-drop microextraction (DDME) and the direct-immersion (DI) as a two-phase system. In its uncomplicated arrangement, one extremely small water droplet which is an unsolvable solvent hanging at the edge of the GC syringe is subjected to the HS of the carried sample or dipped in an aqueous sample (DI mode) in a vial. To increase the rate of extraction process approaches like target compound derivatization, solution salting out, heating, and stirring usually used. When the extraction time is ended, the tiny drop is removed from the syringe. And to conduct instrumental analysis, the enriched solvent is shifted to a selected system without any further operation. The pre-concentration of derivatives or analytes is achieved by HS. The most feasible action for relatively polar compounds and less volatile in pure samples is two-phase approaches.

DDME is an altered version of the DI-SDME technique whose purpose is quantitative preconcentration and swift, economical clean-up of analytes from aqueous solution with least sample utilization. DDME with Chloroform (0.5 ml) was implemented on foods and a single droplet of beverages i.e., 7 ml for the determination of caffeine by Shrivas and Wu (2007). Extraction took place without adding salt at room temperature. It took just 5 min to extract caffeine. By combining gas chromatography-mass spectrometry (GC-MS) with optimized DDME, linearity in the range of 0.05 and 5.0 mg/mL with the limit of detection (LOD) of 4.0 ng/mL, a relative standard deviation (RSD) of 4.4%, recoveries above 97%, the correlation coefficient of 0.980. Major problems of the traditional method of caffeine extraction are lengthy sample pretreatment and a huge quantity of sample and solvent consumption, But DDME handled these problems successfully. Optimized DDME-based procedure is considered as a rapid, simple, and appropriate diagnostic tool for the detection and extraction of caffeine in beverages and food by authors.

Polar compounds are analyzed by using an altered SDME system which is known as LLLME and it is a three-phase SDME system. In that technique, polar analytes that are deionized were collected in an organic phase from the aqueous sample. Later, back-extraction was performed in an aqueous micro-drop which behaves as receiving phase, Uphill now, Biological fluids and aqueous samples are analyzed by that technique. Related to its application of food analysis only one study has come to light. In that study (Zhu et al., 2010), adenine is preconcentrated and online purified from green tea by combining capillary electrophoresis (CE) with LLLME.

3.3.5. HF(2)LPME

HeandLee (1997) proposed the idea of Hollow fiber-protected two-phase liquid mi-cryoextraction (HF(2)LPME) and named it liquid-phase microextraction. In its basic model, the technique includes a hollow fiber polypropylene tube, whose one end is closed, and its diameter is small. It is used to enclose the organic extracting solvent. Fiber is filled with organic solvent for that purpose syringe needle is connected to the open end of the tube. After filling, the tube is placed in the vial which has investigated an aqueous sample. Analyte starts moving through the tube. After a fixed time, the solvent is removed by using the syringe and presented to instruments for analysis usually FOS. The fiber tube is working as a membrane. That's why dirty aqueous samples are preferably analyzed by (HF(2)LPME) instead of SDME. Also, this technique has high stirring rather than SDME because the solvent contained in the hollow fiber has high stability value. On the other hand, the extraction time required for HF-LPME greater than SDME (20–60 min, 5–15 min, respectively).

Moreover, this technique involves higher extraction volumes. During three-phase mode, the small opening of hollow fiber is stuffed with the water-immiscible organic solvent. Once analyte is preconcentrated in the solvent, then it is extracted to an aqueous acceptor phase which implanted in the lumen's fiber. Water-soluble analytes are usually extracted from aqueous matrices by that technique. CE and liquid chromatography (LC) are commonly used for the ultimate instrumental determination of analytes because naturally, extracted analytes are aqueous. In the last years, Many studies have been done to find the suitability of (HF(2)LPME) for using it as an analytical technique for beverages and food-related analytes. There are a number of applications related to this technique Like the analysis of organometallic (Ghasemi et al., 2011) and metallic (Abulhassani et al., 2010) species in complicated foodstuffs, many beverages (Xiong and Hu, 2008); antibiotic in milk (Shariati et al., 2009) and phenolic compounds in fruit juices (Saraji and Mousavi, 2010), orange juice (Barahona et al., 2010) and micropollutants in alcoholic drinks (Plaza Bolanos et al., 2008).

3.3.6. Dispersive liquid-liquid Microextraction

In dispersive liquid-liquid microextraction (DLLME) technique, solution of water-immiscible extraction solvent of a small quantity (averagely, 10–50 mL) dissolved in 0.5–2 mL volume of a water-soluble solvent of range 0.5–2 mL is used to extract investigated aqueous sample (as far as 10 mL). The technique can be regarded as the altered version of LLE. In that technique, extraction is done by small micro-drops of the water-immiscible solvent which is formulated by rapid infusing the combination of organic solvents into the water with the help of a syringe. Freezing or centrifugation (depend upon density) is carried out to remove the enriched organic phase from the aqueous sample. Then, the sample is exposed to instrumental analysis usually to GC without any further processing. In situ derivatization or pH adjustment is desired to implement this technique on polar analytes. This adjustment can be achieved by diffusion together with the extraction solvent or by directly adding a derivatization agent in the sample. It is still updating and miniaturizing since its launch in 2006 (Rezaee et al., 2006). However, its application fields are very wide, and it is impressively manipulative. Comparably abundant food constituents can be determined efficiently and effectively by using that technique as compared to traditional methods, For example, analysis of inorganic (Wen et al., 2011) contaminants, and trace organic (Cunha et al., 2009; Liu et al., 2011b), cholesterol (Daneshfar et al., 2009) and other illegitimate substances (Liu et al., 2011a).

Recently, much research work has been done to utilize ionic liquid as an extractant in HF(2/3) LPME and SDME, the same approach has been developed for DLLME (Table 1.3). Both organic and inorganic compounds are being extracted by using ionic liquids at room temperature instead of traditional organic volatile solvents. This new approach has not only revived the analytical process. It also decreased the risk of harmful solvents for analysts. aqueous samples can directly be subjected to aqueous samples. The target analytes must be extracted from the matrix and extract must be diluted in water to perform that technique on solid matrices. Other than banana classes, eight pesticides have been determined by using that technique by Ravelo-Perez et al. (2009). In this technique, the extraction process provides a fruit sample (1 g) with acetonitrile. After subjecting it to vaporization and making a solution of the extract in water (10 mL), preconcentration of the target compound is achieved by DLLME. In that process, methanol (714 mL) and [HMIM][PF_6] (88 mg) are used as disperser solvent and as extractant respectively. After performing centrifugation at 4000 rpm for 20 min and dilution in acetonitrile, the ionic liquid was regained. It was transferred for analysis avoiding more treatment by LC-DAD. Classic chromatograms produced are presented in Figure 1.4(A) a Spiked (B) a Non-spiked banana. In all instances, LODs (0.32–4.7 mg/kg), RSD values lower than 9%, and acceptable mean recoveries in the 53–97% span were obtained. These are under MRLs fixed in prevailing legislation. The authenticity of the optimized procedure for the desired determination is proved by the above given analytical states. However, the implementation of matrix calibration is mandatory due to the serious matrix effect.

3.3.7. Solid-phase Extraction

Solid-phase extraction (SPE) is the most extensively practiced technique for the analysis of extracts and aqueous samples in the lab. A vast type of sorbents, spanning from traditional sorbents, for example, C8 or C18, florisil, and silica to latest cross-linked polymers are currently easily accessible in many kinds, including traditional SPE disks and cartridges for on-line and ff-line analysis along with that 96-well plates. Numerous reviews written on food analysis include Ridgway et al., 2007; Rostagno et al., 2010); Ihnat, 2003; Kinsellaa et al., 2009; Ihnat, 2003; Kinsellaa et al., 2009; Beyer and Biziuk, 2008; Ihnat, 2003; Kinsellaa et al., 2009;

Buldinia et al., 2002; Ihnat, 2003 and Kinsellaa et al., 2009. Closely relevant research fields, like environmental analysis, are acknowledging the use of SPE for food analysis. Also, universal sorbents are given preference in these researches, i.e., the practice of choosing sorbents depend on molecular imprinted polymers (MIPs) or restricted access media (RAM); the use of deeply cross-linked polymer to enhance the retention of polar analytes; the use of those polymers who can contain both polar and non-polar analytes, so that in a single analysis, number of monitored analytes can be increased (Turiel and Martin-Esteban, 2010). Some major development has been made in nanotechnologies field in the last decade.

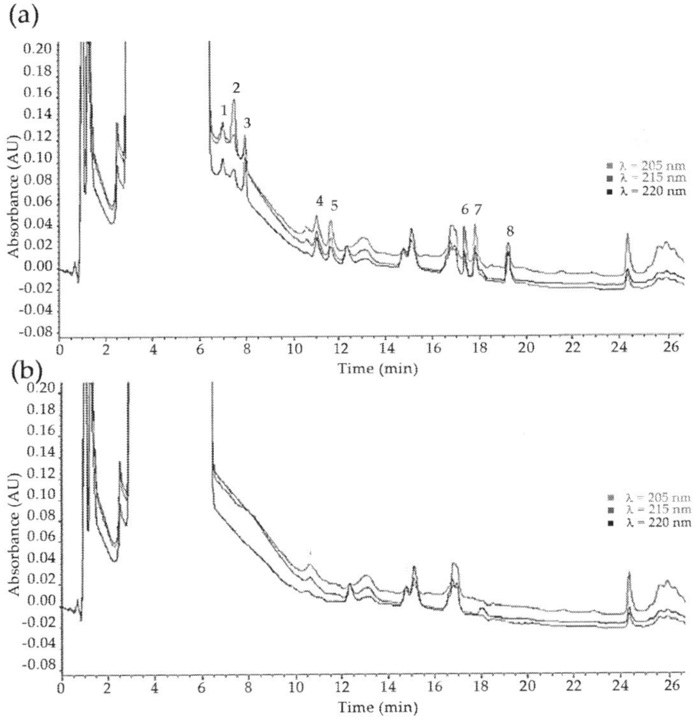

Figure 3.4: After matrix extraction and DLLME with [HMIM][PF6] as extractant, LC-DAD chromatograms of (a) a spiked and (b) non-spiked banana sample. Peak identification: (1) Thiophanate-methyl, (2) carbofuran, (3) carbaryl, (4) tebuconazole, (5) iprodione, (6) oxyfluorfen, (7) hexythiazox, and (8) fenazaquin.

Source: https://www.sciencedirect.com/book/9780123848628/chemical-analysis-of-food-techniques-and-applications.

At present, Food analysis is using the development of nanotechnology, For example, Lopez-Feriaetal presented control of multiclass pesticides in virgin olive oil by utilization of carbon (C) nanotube-based solid-phase extraction (Lopez-Feria et al., 2009). Carboxylated single-walled c nano-tubes (SWCNs) were adopted over Multiwalled C Nanotubes for the operations. Once optimized, diluted investigated olive oil is eluted with n-C_6 (1:5, v/v) by using an SPE column which has selected nanotubes of 30 mg. 3 mL of n-C_6 is used to wash the column. After that, 0.5 mL of ethyl acetate eluted the analytes. Then the concentration of the extract was completed, and methanol was used for reconstruction. Finally, the extract was analyzed by GC-MS. Sample preparation was accomplished not more than 8 min and the column of SPE could be reused for more than 100 times. The target pesticides which are an application of restrictive samples like the ecological virgin olive oil can be controlled because the value of LODs is very low (1.5 to 3.0 mg/L).

QUEChERS has been remembered as one out-standing development in the area of SPE. This method is proposed by Anastassiades et al. (2003) for the analysis of pesticides in vegetables and fruits. The features of that technique are evident by its acronym which is quick, easy, cheap, effective, rugged, and safe. It is a technique that is based on Dispersive solid-phase extraction (DSPE) and has many steps. In (Figure 1.5) the basic methodology for pesticide analysis in vegetables and fruits is shown. The first step is to remove water from the organic solvent with magnesium sulfate treatment. After that, polar components like polar pigments, some sugars, and organic acids are discharged by primary secondary amine (PSA). Other than that, pigments and sterols like chlorophyll can be eliminated by sample shaking. That technique uses graphitized C black (GCB).

The acceptance of that technique is due to its efficiency and rapidness in preparing the samples.

It is being used for other types of analysis. Also, different application, for example, analysis of drugs in the blood (Plossl et al., 2006) and animal tissues (Stubbings and Bigwood, 2009), acrylamide in different food items (Mastovska and Lehotay, 2006) and nonpolar microcontaminants (Ramalhosa et al., 2009).

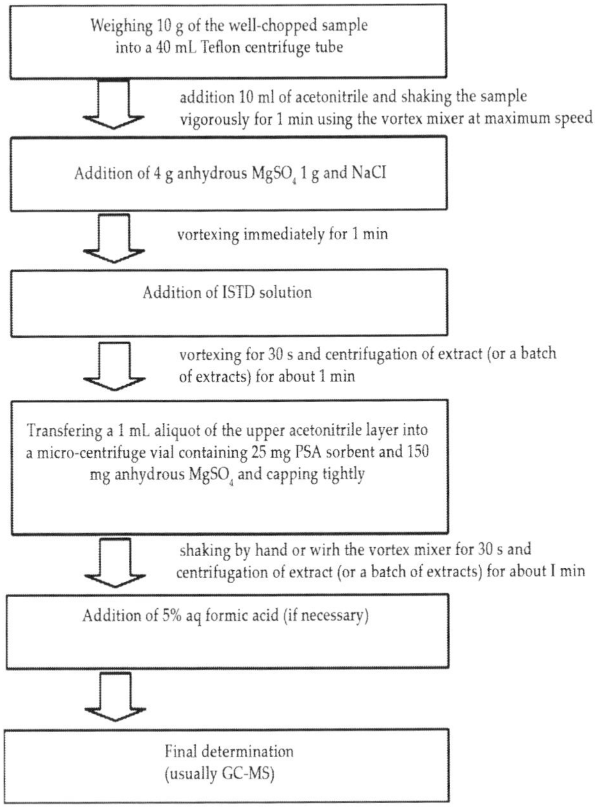

Figure 3.5: Determinization of pesticides in vegetables and fruits using QuECh-ERS technique.

Source: https://www.sciencedirect.com/book/9780123848628/chemical-analysis-of-food-techniques-and-applications.

dSPE has also advantageous because it is used for the development of new materials. To detach tetracycline antibiotics from tissue and egg samples, a magnetic molecularly imprinted polymer has developed using dSPE by Chen et al. (2009). The results acquired by this method is more adequate and enough in contrast to more traditional configurations such as MIP-SPME and MIP-SPE (Table 1.3). This approach is a fascinating substitute for sample preparation due to its simplicity of methodology and ability to retrieve the magnetic flecks with a magnet.

Table 3.3: Comparison of Magnetic MIP with QuEChERS Methods with the Results Derived by Determination of Tetracycline Antibiotics Using MIP-SPME and MIP-SPE

Method	Technique	Sample preparation time	Precision (RSD,%)	Recovery (%)	Reuse of polymer (times)	LOD (ng/g)	Ref.
MIP-SPE	LC-UV	clean-up 1 min for homogeneity þ	<8	66e69	Not mentioned	Not mentioned	Caro et al. (2005)
MIP-SPME	LC-fluorescence	5 or 10 min for homogeneity þ 10 min 30 min for SPME	3e6	72e94	100	1.5e3.5	Hu et al. (2008)
Magnetic MIP	15 or 20 min (simultaneous extraction þ clean-up)	LC-MS/MS	73e96	31e2	0.06e0.19	10	Chen et al. (2009)

3.3.8. Matrix Solid-Phase Dispersion

Samples of solid, semi-solid, viscous, and liquid can comfortably be treated by using a generic technique known as matrix solid-phase dispersion (MSPD). Using that technique, the clean-up and extraction of target analytes are accomplished in one step process and it used format is column format. The column approach is best suitable because it simplified the analytical procedure and unlike conventional LLE methods, it can evade emulsion problems. In the case of foodstuffs, when the extraction solvent protocol and sorbent dispersant are properly selected MSFD can generate such extract which is ready-to-analyze, They are generally handled by GC or LC.The major purpose of MSFD in food analysis is the determination of trace organic micropollutants and, in specific, of pesticides (Kristenson et al., 2006; Gilbert-Lopez et al., 2009; Bogialli and DiCorcia, 2007; Barker, 2007). Sorbents adopted for sample dispersion are from new materials, for example highly selective dedicated sorbents (Yan et al., 2011)or multiwalled C nanotubes (Guan et al., 2011)to traditional sorbents (e.g., C8 or C, Florisil,

Alumina).

A recent tendency related to work on that technique is to use it with one or many described above techniques or to miniaturize the MSPD process (Kristenson et al., 2001; Ramos et al., 2009). The main objective of these trends is to enhance the selectivity of MSPD operation and the overall performance of that technique. In an exemplifying example, Yan et al. (2011) presented the utilization of the latest synthesized type of aniline-naphthol molecularly imprinted microspheres (0.2 g) choose for Sudan as a dispersant for miniaturized MSPD of egg yolk of 0.1 g. MSPD column was washed with a 4 mL mixture of methanol and water in 1:1 by volume (1:1, v/v). After washing, analytes were extracted with 3 mL solution acetone and acetic acid in 95:5 by volume. Concentrated eluent of 1 mL was acted as a dispersive solvent for DLLME. The mixture was stirred, and a homogeneous cloudy solution is generated by ultrasonication. Centrifugation at 4000 rpm for 10 min was carried out to perform phase separation. After the concentration of the enriched phase, the determination process LC-UV was started. In that process, four studied Sudan dyes were detected. The whole sample preparation procedure (Yan et al., 2011) is shown in the schematic diagram Figure 1.6. MSPD achieved proficient linearity for each analyte in the investigated 0.02–2.0 mg/g range ($r \leq 0.9990$). Moreover, RSDs are lesser than 6% and recoveries are more than 87%.

3.3.9. Pressurized Liquid Extraction and Supercritical Fluid Extraction

Pressurized liquid extraction (PLE) and supercritical fluid extraction (SFE) are the techniques which are based on the use of compressed fluids. In these techniques, when water acts like an extractant then it is known as subcritical water extraction (SWE). The purpose of these techniques in food analysis based applications is the separation of functional products and relevant natural compounds (Mendiola et al., 2007).

Extraction of cholesterol from egg and extraction of caffeine from tea and coffee are some of the well-known commercial applications of SFE. Nevertheless, this technique is feasible for several applications due to its peculiar characteristics.

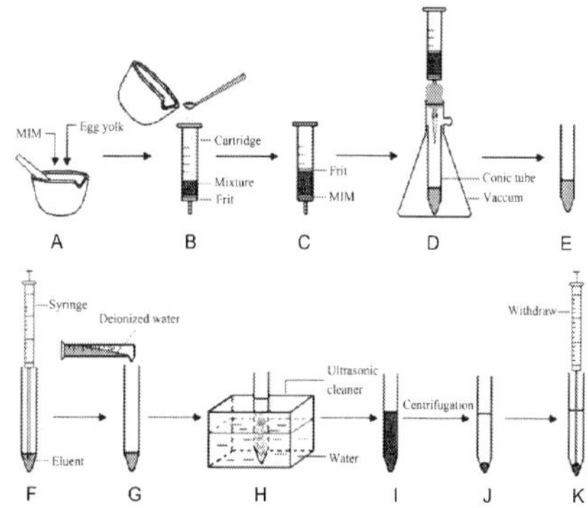

Figure 3.6: Schematic presented for the concurrent determination of four Sudan dyes in egg yolk by the combination of DLLME with MSPD/MIP: (a) Mixing of selective MIP (MIM) with the sample; (b) Pouring of the mixed sample to the column; (c) Test analytes Elution and MSPD column; (e) Evaporation of Eluent, (f) Process of DLLME by injecting the extractant into the eluent; (g) Adding deionized water into the DLLME extractant-dispersant mixture; (h) development of the emulsion with the assistance of ultrasounds; (i) The ternary mixture emulsion; (j)Process of centrifugation for phase separation; and (k) Storage of the high-density extractant.

Source: https://www.sciencedirect.com/book/9780123848628/chemical-analysis-of-food-techniques-and-applications.

The profiles of amino acid of genetically modified soya bean and maize are acquired by applying SFE with a combination of 35% of methanol and C dioxide (CO_2) and combined by GC-MS in offline mode (Bernal et al., 2008; Zougagh and Rios, 2008). The observation is made in between these profiles and the profiles acquired from their corresponding isogenic nontransgenic varieties and it is evident that isogenic nontransgenic varieties have excessive content of amino acids.

The features of SFE that make it special and widely used technique are its comparatively simple and easy on-line or at-line joining with other separation-plus-detection instruments and its usage in automatic sample treatment (Pol et al., 2004; Carabias-Martinez et al., 2005; Mendiola et al., 2007).

REFERENCES

1. 2002/63/EC, C. d, (2002). Commission directive 2002/63/EC of 11 July 2002 establishing Community methods of sampling for the official control of pesticides in and on products of plants and animal origin and repealing Directive 70/700/EEC. *Official Journal of European Communities L, 187*, 30–43.
2. 90/642/EEC,D,(1993).AnnexItoDirective90/642/EEC[Partofproducts to which maximum limits apply], as mentionedbyDirective93/58/EEC (OJL211), 23.8.1993, pp. 1–26.
3. Abulhassani, J., Manzoori, J. L., and Amjadi, M., (2010). Hollow fiber based-liquid phase microextraction using ionic liquid solvent for preconcentration of lead and nickel from environmental and biological samples prior to determination by electrothermal atomic absorption spectrometry. *Journal of Hazardous Materials, 176*, 481–486.
4. Anastassiades, M., Lehotay, S., Stajnbaher, D., and Schenk, F., (2003). Fast and easy multi-residue method employing acetonitrile extraction/partitioning and "dispersive solid-phase extraction" for the determination of pesticide residues in produce. *Journal of AOAC International, 86*, 412–431.
5. AOAC, (1990). In: Helrich, K., (ed.), *Official Methods of Analysis of the AOAC*. AOAC, USA (Vol. 1, pp. 1-33).
6. Arthur, C. L., Pawliszyn, J., (1990). Solid phase micro-extraction with thermal desorption using fused silica optical fibers. *Analytical Chemistry, 62*, 2145–2148.
7. Asensio-Ramos, M., Ravelo-Perez, L. M., Gonzalez-Curbelo, M. A., and Herna´ndez-Borges, J., (2011). Liquid phase microextraction applications in food analysis. *Journal of Chromatography, 1*, 1–30. doi:10.1016/j.chroma. 2011.5.96.
8. Barahona, F., Gjelstad, A., Pedersen-Bjergaard, S., and Rasmussen, K. E., (2010). Hollow fiber-liquid-phase micro extraction of fungicides from orange juices. *Journal of Chromatography A, 1217*, 1989–1994.
9. Barker, S. A., (2007). Matrix solid phase dispersion (MSPD). *Journal of Biochemical and Biophysical Methods 70*, 151–162.
10. Bernal, J. L., Nozal, M. J., Toribio, L., Diego, C., Mayo, R., and Maestre, R., (2008). Use of supercritical fluid extraction and gas chromatography-mass spectrometry to obtain amino acid profiles from several genetically modified varieties of maize and soybean. *Journal of*

Chromatography A, 1192, 266–272.

11. Beyer, A., and Biziuk, M., (2008). Applications of sample preparation techniques in the analysis of pesticides and PCBs in food. *Food Chemistry 108*, 669–680.
12. Bogialli, S., and Di Corcia, A., (2007). Matrix solid-phase dispersion as a valuable tool for extracting contaminants from foodstuffs. *Journal of Biochemical and Biophysical Methods 70*, 163–179.
13. Buldinia, P. L., Riccib, L., and Sharma, J. L., (2002). Recent applications of sample preparation techniques in food analysis. *Journal of Chromatography A, 975*, 47–70.
14. Carabias-Martinez, R., Rodriguez-Gonzalo, E., Revilla-Ruiz, P., and Hernandez-Mendez, J., (2005). Pressurized liquid extraction in the analysis of food and biological samples. *Journal of Chromatography A, 1089*, 1–17.
15. Caro, E., Marce, R. M., Cormack, P. A. G., Sherrington, D. C., and Borrull, F., (2005). Synthesis and application of an oxytetracycline imprinted polymer for the solid-phase extraction of tetracycline antibiotics. *Analytica Chimica Acta, 552*, 81–86.
16. Chen, L., Liu, J., Zeng, Q., Wang, H., Yu, A., and Zhang, H., Ding, L., (2009). Preparation of magnetic molecularly imprinted polymer for the separation of tetracycline antibiotics from egg and tissue samples. *Journal of Chromatography A, 1216*, 3710–3719.
17. Cunha, S. C., Fernandes, J. O., Oliveira, and M. B. P. P., (2009). Fast analysis of multiple pesticide residues in apple juice using dispersive liquid-liquid micro extraction and multidimensional gas chromatography-mass spectrometry. *Journal of Chromatography A, 1216*, 8835–8844.
18. Curren, M. S. S., and King, J. W., (2002). Sampling and sample preparation for food analysis. In: Pawliszyn, J., (ed.), Sampling, and sample preparation for field and laboratory. Fundamentals and new directions in sample preparation. *Comprehensive Analytical Chemistry* (Vol. XXXVII, pp. 1–50). Elsevier, Amsterdam.
19. Daneshfar, A., Khezeli, T., and Lotfi, H. J., (2009). Determination of cholesterol in food samples using dispersive liquid-liquid micro extraction followed by HPLC-UV. *Journal of Chromatography B, 877*, 456–460.
20. Fan, Z., and Zhou, W., (2006). Dithizone-chloroform single drop

microextraction system combined with electrothermal atomic absorption spectrometry using Ir as permanent modifier for the determination of Cd in water and biological samples. *Spectrochimica Acta Part B: Atomic Spectroscopy, 61*, 870–874.

21. Gao, S., You, J., Zheng, X., Wang, Y., Ren, R., Zhang, R., Bai, Y., and Zhang, H., (2010). Determination of phenylurea and triazine herbicides in milk by microwave assisted ionic liquid micro extraction high-performance liquid chromatography. *Talanta 82*, 1371–1377.
22. Ghasemi, E., Sillanpaa, M., and Najafi, N. M., (2011). Headspace hollow fiber protected liquid-phase microextraction combined with gas chromatography-mass spectroscopy for speciation and determination of volatile organic compounds of selenium in environmental and biological samples. *Journal of Chromatography A, 1218*, 380–386.
23. Gilbert-Lopez, B., Garcia-Reyes, J. F., and Molina-Diaz, A., (2009). Sample treatment and determination of pesticide residues in fatty vegetable matrices: A review. *Talanta 79*, 109–128.
24. Guan, S., Yu, Z., Yu, H., Song, C., Song, Z., and Qin, Z., (2011). Multi-walled carbon nanotubes as matrix solid-phase dispersion extraction adsorbent for simultaneous analysis of residues of nine organophosphorus pesticides in fruit and vegetables by rapid resolution LC-MS-MS. *Chromatographia 73*, 33–41.
25. He, Y., and Lee, H. K., (1997). Liquid-phase micro extraction in a single drop of organic solvent by using a conventional micro syringe. *Analytical Chemistry, 69*, 4634–4640.
26. Hu, X., Pan, J., Hu, Y., Huo, Y., and Li, G., (2008). Preparation and evaluation of solid-phase micro extraction fiber based on molecularly imprinted polymers for trace analysis of tetracycline's in complicated samples. *Journal of Chromatography A, 1188*, 97–107.
27. Ihnat, M., (2003). Sample preparation for food analysis. *Comprehensive Analytical Chemistry 41*, 765–856.
28. Kinsellaa, B., O'Mahonya, J., Malonec, E., Moloneya, M., Cantwella, H., Fureyb, A., and Danaher, M., (2009). Current trends in sample preparation for growth promoter and veterinary drug residue analysis. *Journal of Chromatography A, 1216*, 7977–8015.
29. Kristenson, E. M., Haverkate, E. G. J., Slooten, C. J., Ramos, L., Vreuls, R. J. J., and Brinkman, U. A. T., (2001). Miniaturized automated matrix solid-phase dispersion extraction of pesticides in fruit followed

by gas chromatographic-mass spectrometric analysis. *Journal of Chromatography A, 917*, 277–286.

30. Kristenson, E. M., Ramos, L., and Brinkman, U. A. T., (2006). Recent advances in matrix solid-phase dispersion. *Trends in Analytical Chemistry, 25*, 96–111.
31. Leon, C., Rodriguez-Meizoso, I., Lucio, M., Garcia-Canas, V., Ibanez, E., Schmitt-Kopplinb, P., and Cifuentes, A., (2009). Metabolomics of transgenic maize combining Fourier transform-ion cyclotron resonance-mass spectrometry, capillary electrophoresis-mass spectrometry and pressurized liquid extraction. *Journal of Chromatography A, 1216*, 7314–7323.
32. Liu, B., Yan, H., Qiao, F., and Geng, Y., (2011a). Determination of clenbuterol in porcine tissues using solid-phase extraction combined with ultrasound-assisted dispersive liquid-liquid microextraction and HPLC-UV detection. *Journal of Chromatography B, 879*, 90–94.
33. Liu, X., Zhao, A., Zhang, A., Liu, H., Xiao, W., Wang, C., and Wang, X., (2011b). Dispersive liquid-liquid micro-extraction and gas chromatography-mass spectrometry determination of polychlorinated biphenyls and poly-brominated diphenyl ethers in milk. *Journal of Separation Science 34*, 1084–1090.
34. Lopez-Feria, S., Cardenas, S., and Valcarcel, M., (2009). One-step carbon nanotubes-based solid-phase extraction for the gas chromatographic-mass spectrometric multiclass pesticide control in virgin olive oils. *Journal of Chromatography A, 1216*, 7346–7350.
35. Manzoori, J. L., Amjadi, M., and Abulhassani, J., (2009). Ultra-trace determination of lead in water and food samples by using ionic liquid-based single drop micro extraction-electro-thermal atomic absorption spectrometry. *Analytical Chimica. Acta, 644*, 48–52.
36. Martendal, E., Budziak, D., and Carasek, E., (2007). Application of fractional factorial experimental and Box-Behnken designs for optimization of single-drop micro extraction of 2,4,6-trichloroanisole and 2,4,6-tribromoanisole from wine samples. *Journal of Chromatography A, 1148*, 131–136.
37. Mastovska, K., and Lehotay, S. J., (2006). Rapid sample preparation method for LC-MS/MS or GC-MS analysis of acryl-amide in various food matrices. *Journal of Agricultural and Food Chemistry, 54*, 7001–7008.

38. Mendiola, J. A., Herrero, M., Cifuentes, A., and Ibanez, E., (2007). Use of compressed fluids for sample preparation: Food applications. *Journal of Chromatography A, 1152*, 234–246.
39. Mol, H. G. J., Rooseboom, A., Van, D. R., Roding, M., Arondeus, K., and Sunarto, S., (2007). Modification and re-validation of the ethyl acetate-based multi-residue method for pesticides in produce. *Analytica and Bio-analytical Chemistry 389*, 1715–1754.
40. Nollet, L. M. L., (1996). In: Nollet, L. M. L., (ed.), *Handbook of Food Analysis*. Marcel Dekker, USA.*1*, 1-44.
41. Olariu, R. I., Vione, D., Grinberg, N., and Arsene, C., (2010). Sample preparation for trace analysis by chromato-graphic methods. *Journal of Liquid Chromatography and Related Technologies, 33*, 1174–1207.
42. Plaza Bolanos, P., Romero-Gonzalez, R., GarridoFrenich, A., and Martinez Vidal, J. L., (2008). Application of hollow fiber liquid phase micro extraction for the multi residue determination of pesticides in alcoholic beverages by ultra-high pressure liquid chromatography coupled to tandem mass spectrometry. *Journal of Chromatography A, 1208*, 16–24.
43. Plossl, F., Giera, M., and Bracher, F., (2006). Multiresidue analytical method using dispersive solid-phase extraction and gas chromatography/ ion trap mass spectrometry to determine pharmaceuticals in whole blood. *Journal of Chromatography A, 1135*, 19–26.
44. Pol, J., Hyotylainen, T., Ranta-Aho, O., and Riekkola, M. L., (2004). Determination of lycopene in food by on-line SFE coupled to HPLC using a single monolithic column for trapping and separation. *Journal of Chromatography A, 1052*, 25–31.
45. Raitioa, R., Orliena, V., and Skibsted, L. H., (2011). Storage stability of cauliflower soup powder: The effect of lipid oxidation and protein degradation reactions. *Food Chemistry, 128*, 371–379.
46. Ramalhosa, M. J., Paıga, P., Morais, S., Delerue-Matos, C., and Oliveira, M. B., (2009). Analysis of polycyclic aromatic hydrocarbons in fish: evaluation of a quick, easy, cheap, effective, rugged, and safe extraction method. *Journal of Separation Science, 32*, 3529–3538.
47. Ramos, J. J., Dietz, C., Gonzalez, M. J., and Ramos, L., (2007). Miniaturized selective pressurized liquid extraction of polychlorinated biphenyls. *Journal of Chromatography A, 1152*, 254–261.
48. Ramos, J. J., Gonzalez, M. J., and Ramos, L., (2009). Comparison of

gas-chromatography-based approaches after fast miniaturized sample preparation for the monitoring of selected pesticide classes. *Journal of Chromatography A, 1216*, 7307–7313.

49. Ravelo-Perez, L. M., Hernandez-Borges, J., Asensio-Ramos, M., and Rodriguez-Delgado, M. A., (2009). Ionic liquid based dispersive liquid-liquid microextraction for the extraction of pesticides from bananas. *Journal of Chromatography A, 1216*, 7336–7345.
50. Rezaee, M., Assadi, Y., Milani-Hosseini, M. R., Aghaee, E., Ahmadi, F., and Berijani, S., (2006). Determination of organic compounds in water using dispersive liquid-liquid microextraction. *Journal of Chromatography A, 1116*, 1–9.
51. Ridgway, K., Lalljie, S. P. D., and Smith, R. M., (2007). Sample preparation techniques for the determination of trace residues and contaminants in foods. *Journal of Chromatography A, 1153*, 36–53.
52. Ridgway, K., Lalljie, S. P. D., and Smith, R., (2010). An alternative method for analysis of food taints using stir bar sorptive extraction. *Analytica. Chimica. Acta, 677*, 29–36.
53. Rocha, S. M., Coelho, E., Zrostlikova, J., Delgadillo, I., and Coimbra, M. A., (2007). Comprehensive two-dimensional gas chromatography with time-of-flight mass spectrometry of monoterpenoids as a powerful tool for grape origin traceability. *Journal of Chromatography A, 1161*, 292–299.
54. Rostagno, M. A., DArrigo, M., and Martinez, J. A., (2010). Combinatory and hyphenated sample preparation for the determination of bioactive compounds in foods. *TrAC Trends in Analytical Chemistry, 29*, 553–561.
55. Saraji, M., and Mousavi, F., (2010). Use of hollow fiber-based liquid-liquid-liquid microextraction and high-performance liquid chromatography-diode array detection for the determination of phenolic acids in fruit juices. *Food Chemistry, 123*, 1310–1317.
56. Saraji, M., and Mousavinia, F., (2006). Single-drop microextraction followed by in-syringe derivatization and gas chromatography-mass spectrometric detection for determination of organic acids in fruits and fruit juices. *Journal of Separation Science, 29*, 1223–1229.
57. Shariati, S., Yamini, Y., and Esrafili, A., (2009). Carrier mediated hollow fiber liquid phase microextraction combined with HPLC-UV for preconcentration and determination of some tetracycline antibiotics.

Journal of Chromatography B, 877, 393–400.

58. Shrivas, K., and Wu, H. F., (2007). Rapid determination of caffeinexe caffeine in one drop of beverages and foods using drop-to-drop solvent microextraction with gas chromatography/mass spectrometry. *Journal of Chromatography A, 1170*, 9–14.
59. Shrivas, K., and Wu, H. F., (2008). Ultrasonication followed by single-drop microextraction combined with GC/MS for rapid determination of organochlorine pesticides from fish. *Journal of Separation Science, 31*, 380–386.
60. Stubbings, G., and Bigwood, T., (2009). The development and validation of a multiclass liquid chromatography tandem mass spectrometry (LC-MS/MS) procedure for the determination of veterinary drug residues in animal tissue using a QuEChERS (QUick, Easy, CHeap, Effective, Rugged, and Safe) approach. *Analytica, Chimica, Acta, 637*, 68–78.
61. Tankeviciute, A., Kazlauskas, R., and Vickackaite, V., (2001). Headspace extraction of alcohols into a single drop. Analyst *126*, 1674–1677.
62. Turiel, E., and Martin-Esteban, A., (2010). Molecularly imprinted polymers for sample preparation: a review. *Analytica, Chimica, Acta, 668*, 87–99.
63. Wang, L., Wang, Z., Zhang, H., Li, X., and Zhang, H., (2009). Ultrasonic nebulization extraction coupled with headspace single drop microextraction and gas chromatography-mass spectrometry for analysis of the essential oil in *Cuminumcyminum L. Analytica. Chimica. Acta, 647*, 72.
64. Wen, X., Yang, W., Yan, Z., and Deng, Q., (2011). Determination of cadmium and copper in water and food samples by dispersive liquid-liquid microextraction combined with UV-vis spectrophotometry. *Microchemical. Journal 97*, 249–254.
65. Wilkowska, A., and Biziuk, M., (2011). Determination of pesticide residues in food matrices using the QuEChERS methodology. *Food Chemistry, 125*, 803–812.
66. Woodget, B. W., and Cooper, D., (1987). *Samples and Standards* (Vol. 1, pp. 2–30). John Wiley and Sons, Chichester.
67. Xiao, Q., Hu, B., and He, M., (2008). Speciation of butyl tin compounds in environmental and biological samples using headspace single drop microextraction coupled with gas chromatography-inductively coupled

plasma mass spectrometry. *Journal of Chromatography A, 1211*, 135–141.

68. Xiong, J., and Hu, B., (2008). Comparison of hollow fiber liquid phase microextraction and dispersive liquid-liquid microextraction for the determination of organosulfur pesticides in environmental and beverage samples by gas chromatography with flame photometric detection. *Journal of Chromatography A, 1193*, 7–18.
69. Yan, H., Wang, H., Qiao, J., and Yang, G., (2011). Molecularly imprinted matrix solid-phase dispersion combined with dispersive liquid-liquid microextraction for the determination of four Sudan dyes in egg yolk. *Journal of Chromatography A, 1218*, 2182–2188.
70. Zhao, E., Han, L., Jiang, S., Wang, Q., and Zhou, Z., (2006). Application of a single-drop microextraction for the analysis of organophosphorus pesticides in juice. *Journal of Chromatography A, 1114*, 269–273.
71. Zhu, Z., Zhou, X., Yan, N., Zhou, L., and Chen, X., (2010). On-line combination of single-drop liquid-liquid-liquid micro-extraction with capillary electrophoresis for sample cleanup and preconcentration: A simple and efficient approach to determining trace analyte in real matrices. *Journal of Chromatography A, 1217*, 1856–1861.
72. Zougagh, M., and Rios, A., (2008). Supercritical fluid extraction as an on-line clean-up technique for determination of riboflavin vitamins in food samples by capillary electrophoresis with fluorimetric detection. *Electrophoresis, 29*, 3213–3219.

Chapter 4

Water Characteristics in Food Technologies

CONTENTS

4.1. INTRODUCTION

On Earth, water is a unique substance that occurs simultaneously in all three physical conditions (liquid/water, gas/water vapor, and solid/ice). It is an important substance for life, i.e., living systems comprise of water and its quantity is about 70 to 80% (w/w). Water owns many physical and chemical properties that make it beneficial to organisms and cells. Water acts as a medium for a number of biochemical reactions (Hui et al., 2006).

Water is a key nutrient in nearly all the food groups including meats, dairy products, vegetables, fruits, and grains. Main nutrients like water-soluble vitamins, proteins, minerals, and carbohydrates are hydrophilic. In foods, most of proteins and carbohydrates are usually plasticized by water. A portion of lipid constituents is hydrophilic, while another part of protein and lipid constituents are hydrophobic and suffer from hydrophilic influence in an aqueous medium (Lerner and Lerner, 2006; Fennema et al., 2017).

Water interrelates with other food constituents using hydrogen-bonding, hydrophobic interactions, and polar interactions. These interactions alter the properties of water. Water plays numerous essential roles within the field of food, feed science, technology, agriculture, and engineering. Water is the key component of different beverages and the majority of the foodstuffs. The water content of vegetables, seafood, meats, and fresh fruits surpass 50%. To cook foods, water is utilized as a good medium. Moreover steam is the essential working medium of power engineering and heat (as a feed for boiler) in food processing (Roos, 1993; Robinson and Eskin, 2000). In food chains, water isn't just a medium for the reactions but is an active component used to control food texture, reactions, biological, and physical behavior. Food might lose moisture to the environment or pick up moisture from the environment during storage. By controlling the activity of water during storage, the loss percent of fresh foods can be considerably minimized. These alterations might impact the organoleptic properties and texture of foodstuffs. Many procedures in food processing (dehydration, thawing, drying, concentration, freezing, and salting) use state transition properties and physicochemical properties of water (Belitz et al., 2009; Alais, 2012).

In the literature, substantial amounts of information on water properties are available. The information and its significance to food technology are fragmented and dispersed. The purpose of this chapter is to define numerous properties of water, to discuss the influence of water properties on combined foods and individual food constituents, to describe its significance to food technology, and to associate water properties with corresponding properties of similar substances. (Lewicki, 2004; Lide, 2004). Numerous properties of water (heat of fusion, heat capacity, surface tension, viscosity, the heat of vaporization and sublimation, boiling, and melting points and thermal conductivity) have been related to similar substances. Data provided in this chapter can be used for storage, packaging, and processing of foodstuffs and beverages. (DeMan et al., 1999; Hardy et al., 2002).

4.2. VISCOSITY

Thermal conductivity and viscosity are two of the main transport properties. Dynamic viscosity of water is normally greater as compared to that of similar substances without hydrogen bonds. The viscosity declines intensely with an increment in temperature. The dependence generally can be defined by a conventional Arrhenius type expression (Kirillin, 1986; Blond and Le Meste, 2004).

Aqueous solutions and pure water comprising low molecular weight (MW) solutes like sugars and salts under normal conditions are the Newtonian fluids (shear stress is directly related to the rate of shear strain or rate of deformation). There is an increase in viscosity when a nonvolatile material is dissolved in water. Generally, aqueous solutions are comprised of large MW food constituents (polysaccharides, proteins, hydrocolloids, and polysaccharides), although only small quantities are non-Newtonian (Engberts, 2007).

4.3. SURFACE TENSION

With the exemption of mercury, water has the utmost surface tension as compared to any other liquids. Non-Polar compounds incline to have much lower surface tension than the polar compounds. The surface tension of water is considerably higher (72.8 mN/mat 20°C) than ethyl alcohol. Hydrogen bonding of the water molecules is liable for its higher surface tension value. Hydrogen bonding is credited to the water's ability to wet to or adhere to most surfaces (hydrophilic substances). Other substances like waxes, fats, oils, and several synthetics polymers do not wet with water (Chen, 2009). There is an association between hydrophilicity of food constituent along with surface hydrophobicity, like protein and its capacity to change interfacial and surface tensions. The surface tension of water reduces with a rise in temperature; the decrease shows approximately linear behavior as a function of its absolute temperature. At the critical temperature, the surface tension of water is zero. In the food industry, disodium hydrogen phosphate and dioctyl sodium (Na+) sulfosuccinate have been utilized as surface tension regulators (Roos, 2002; Franks, 2007).

4.4. THE SOLUBILITY OF MATERIALS IN WATER

The thermodynamic condition of water (illustrated by its Gibbs free energy of water which controls the feasibility of physical or chemical transformation) in food surfaces first from rare properties of water and its capability to make strong hydrogen bonds (Van Oss, 2008). The materials porosity, which is relatively common for foods, also disturbs the thermodynamic condition of the water. Interrelation with a hydrophilic solute triggers the creation of the hydrating components. Conversely, interrelations with hydrophobic solute greatly affect the structural orientation of the solvent water (Chieh, 2005). The dissolution of certain food constituents like sugars in water is due to bonding amongst polar groups of food constituents and water molecules. The dissolution of salts is because of electrostatic forces among water and ions produced from the salts. The water molecule significantly weakens hydrogen bonds and electrostatic forces among polar molecules by contending with the attraction forces. The forces of hydrogen bonding present in water molecules are larger than the attraction forces among ions of salts (Slade et al., 1991; Isengard, 2001; Da Pieve et al., 2010).

Water is utilized as diluents or solvent in the preparation of foodstuffs and beverages. A foodstuff normally contains several inorganic and organic substances. These solutes alter the properties of water. At ambient pressure and temperature, the food components (minerals, acids, carbohydrates, and proteins) are miscible in water, whereas lipids have very little interactions with water (Lewicki, 2009; Rahman and Sablani, 2009).

Water is categorized as hard, medium, and soft. Water hardness is categorized based on the quantities of detachable salt calcium carbonate per unit volume from water. In food processing, water hardness is the main factor. It can intensely affect the quality of food products. Water hardness might be treated or altered by utilizing a chemical ion exchange system. Water hardness is influenced by the pH of the medium, which plays an acute role in food processing. For instance, hard water averts the successful manufacturing of clear beverages (Dahl and Andersen, 1983).

Physical properties of the water (e.g., freezing, and boiling points) are usually affected by solutes like as sugars and salts in water, foodstuffs, and beverages. One mole of Na+ chloride per kilogram of water increases the boiling point of water by 1.02°C and 1 mole of sucrose per kg increases the boiling point by 0.51°C. Likewise, enhancing a number of the dissolved particles decreases the freezing point of water. For instance, it is −16 and −8.4°C for 7% solutions of NaCl and K_2SO_4 respectively. Freezing point depression of water is interrelated to the water activity (Atkins et al., 2018). Freezing point depression is utilized for milk contaminations checking (Rockland and Beuchat, 1987; Coultate, 2009). This water property can also be utilized to confirm for contamination of food and other beverages. Extraction, filtration, adsorption, and osmotic processes can be used to remove numerous contaminants from both surface and groundwater. When salty water freezes drinking water might be prepared. The drinking water formed from the last procedure is nearly pure, and it includes very little salts (Suresh and Naik, 2000; Lewicki, 2009).

4.5. WATER ACTIVITY AND WATER CONTENT

Water, as a component of food, might affect food quality, stability, safety, and physical properties. In a food variety of water, concentration is very wide and initiates with a fraction of a percent and extends even more than 98%. Liquid and fresh foods contain normally large amounts of water whereas dried and baked foods are very poor in water content (Eubanks, 2006; Stanga, 2010). Food products having large water content, like fresh foods, water gain triggers to satisfactory conditions and water loss changes physical characteristics for microbiological progression. Therefore, food packaging is done against a loss or gain of water (Fox and McDonald, 1994; Chaplin, 2006).

Water activity is termed as a ratio of the vapor pressure the water (H_2O) in a solution to the vapor pressure of the pure water. Water activity, at any water content and temperature, can be calculated by determining the equilibrium relative humidity (ERH) covering the food in a closed container. It is an estimate of available water or free moisture in the food product (Cybulska and Doe, 2006; Kasaai, 2008). Water activity is utilized to evaluate whether foodstuff losses or absorbs water in a certain environment. It is a critical limit to estimate food quality. The water activity in foodstuff or the solution is originated from its chemical potential. The pure water chemical potential is given by molar Gibbs free energy. The water activity shows the combined effects of water-surface, capillary, water-solute, hydrophobic, and hydrophilic interactions (Zemansky et al., 1975; Gun'ko et al., 2005).

Several food properties relate in a better way with water activity than with the water content, specifically if the food includes substances that are less or more indifferent to water, like oil droplets or crystals (Eisenberg and Kauzmann, 2005). This doesn't mean that associations amongst water activity and property are similar for all foods, not even in the comparative sense. Some properties like notability rheological ones, usually relate in a better way with water content than with the water activity. So, water content knowledge isn't sufficient information to evaluate food stability and safety during storage (Busk Jr, 1984; Kasaai, 2010).

Water content as the function of vapor pressure at the stability (P/P_0) and constant temperature is nominated as absorption isotherm. Water inclines to be adsorbed in the shape of clusters. Generally, a monolayer of water, Brunauer-Emmert-Teller (BET) is severe water content, and dehydrated foods are steady below this. The BET isotherm is appropriate at low water activity. A plot of humidity content vs. water activity at any temperature can be utilized: (i) to optimize dehydration, concentration, and formulation, processes; (ii) to enhance packaging conditions and to regulate microbial growth; and (iii) to control chemical and physical stability of food products. When the value of water activity is low, foods (proteins, lipids, carbohydrates) act more like rubbery polymers than the crystalline structures. The water substance might be physically halted as the glassy material, it might be bound to the food constituents (carbohydrates or proteins) in food methods with very low a_w or it might be bound with food constituents under freezing circumstances. Under the last conditions, the molecule of water might diffuse with trouble and therefore might hinder the diffusion of solutes. Alterations in the structure of proteins and carbohydrates from crystalline to amorphous or vice versa might also occur very slowly by water diffusion or migration (Kireev et al., 1972; Sammon et al., 2003). Information on water absorption performance, and its association to the glass transition temperature, T_g gives a valuable tool: (a) to regulate material behavior during subsequent storage and processing; and (b) to confirm food packaging requirements. The T_g is a significant parameter for long tenure storage steadiness of frozen foods (e.g., ice creams) and monitoring textural properties of chips and cookies. For dried and baked foods, which are kept at room temperature, the T_g may specify a temperature at which loss of softness, growth of an undesirable texture, staling, may take place (Hatakeyama and Hatakeyama, 1998).

The water activity of liquid foods and solutions depends on the chemical nature of solute, temperature, and concentration. The water activity of fresh foods (most dairy products, seafood, vegetable, meats, and fresh fruits) is close to 1 and high. Moreover, products having high fat with comparatively small amounts of water, such as margarine and butter, have a comparatively high water activity. When water activity is lesser than 0.9, many molds are withdrawn. Most of the microorganisms stop growing if water activity is lesser than 0.6. The decrease of water activity delays the growth of microorganisms, enzymatic reactions and non-enzymatic browning (Labuza, 1977; Coles et al., 2003).

The amount of desorption of foods and water adsorption depends on the temperature and physical states. The main physical change that happens during storage of the frozen food outcomes from water migration. The water migration affects water content alterations or changes in the size of the ice crystal called re-crystallization. Water migration can relate either to changing the water place for the products with mixed water contents or to the water loss by surface dehydration (Gun'ko and Turov, 1999; Mathlouthi, 2013).Water activity is an important parameter, as it affects taste, safety, appearance, texture, and shelf foods life. Water activity is a significant factor in dehydrated foods and concentrated foods or foods that include sugars and salts (Crosby, 1981; Gallier et al., 1987). Frequently it is of great concern in the development of food products to estimate the result of food composition on the water activity of final products. It is also significant in the packaging of foodstuffs with changed ingredients having altered water activities. A foodstuff comprising several constituents with altered water activities are dependent on time for the water migration until equilibrium is achieved. The rate of deterioration and the level of water loss depend intensely on water activity and water content (Huffman et al., 1986; Li and Ross, 1993).

4.6. THE REACTION OF WATER WITH OTHER COMPONENTS OF FOOD

Numerous reactions, like as biochemical or chemical hydrolysis (proteins, carbohydrates, lipids) and non-enzymatic and enzymatic (Maillard) browning, happen in the existence of water. The oxidative reaction is a normal reaction in the low moisture foods. Water can be oxidizing or reducing reagent because it gives electrons or protons. Oxidation-reduction reactions comprise water generally due to electron or proton transfer. Oxidation-reduction reactions of water give erosion on the metal surfaces. The later reactions in the food industry consequence in erosion in reaction vessels and processing lines and generate metal ions (Fe^{+2}, Pb^{2+}, and Cu^{2+}) in beverages, foodstuffs, and drinking water. The occurrence of certain metal ions (Sn^{+2}, Fe^{+2}, Pb^{2+}, Cu^{2+}) as remains in canned foods, beverages, and drinking water, the outcome in off-flavor and unsafe products (Edsall and McKenzie, 1983; Jenkins and Morrison, 1999).The reactions of food constituents along with rates of many reactions in food chains are disturbed by water activity. The food stability, rates of numerous deteriorative reactions (non-enzymatic and enzymatic browning, enzyme-catalyzed flavor changes, biochemical or chemical hydrolysis, oxidative reactions, anthocyanin degradation, and chlorophyll degradation) and microbial development depends on the water content, state of water and water activity (Wiggins, 1971). The rates of mutual non-enzymatic and enzymatic reactions enhance as water activity enhances. The decline of vitamin C enhances rapidly as water activity enhances. Rates of nearly all reactions enhance as water content enhances over the lower part of a multilayer zone. Adding the water outside the boundary of multilayer and mono-layers zones results in an enhancement in the rates of oxidation. Yong and Karel (1981) have recommended that adding the water in this area of the isotherm may boost oxidation by enhancing the solubility of oxygen and through permitting macromolecules to swell, thus exposing more catalytic places. In fat comprising foods and dried with very small moisture substances oxidation proceeds very speedily. At water activity equal or lesser to 0.3, lipid oxidation delays by decreasing catalytic activity of the metal catalysts, by hindering the access of oxygen to the foods and/or promoting non-enzymatic browning. At upper water activities (a_w = 0.55 to 0.80), the oxidation rate enhances by enhancing catalysts mobilization. The water activities extra than 0.80, the additional water may delay rates of oxidation, and dilution of catalysts decreases their effectiveness (Le Bail, 2004; Jiang and Lee, 2006).The amounts of non-enzymatic browning are highest in foods with intermediary moisture matters (roasted nuts, toasted breakfast cereals, roller-dried milk powder, and bakery products). Disagreeable Maillard reaction can be reserved by reducing moisture to a very small level or by enhancing dilution (Liu and De Yao, 2001; Welti et al., 2004).By concentration of solutes, the reaction rates are influenced. Elimination of water from the foods in the procedure of dehydration, freezing, and concentration of foods decreases the rate of deterioration (enzymatic and microbial). The reduction in temperature usually reduces the rate of biochemical and chemical reactions. Deteriorative reactions are braked at 0°C and the usual temperature of ice-covered storage (−18°C).

REFERENCES

1. Alais, C., (2012). *Food Biochemistry* (Vol. 1, pp. 1–30). Springer Science and Business Media.
2. Atkins, P. W., De Paula, J., and Keeler, J., (2018). *Atkins' Physical Chemistry* (Vol. 1, pp. 10–30). Oxford university press.
3. Belitz, H. D., Grosch, W., and Schieberle, P., (2009). Food additives. *Food Chemistry*, *1*, 429–466.
4. Blond, G., and Le Meste, M., (2004). Principles of frozen storage. *Food Science and Technology* (Vol. 1, pp. 25–54). New York-Marcel Dekker.
5. Busk, Jr. G. C., (1984). Polymer-water interactions in gelation. *Food Technology (USA)*, *1*, 20–40.
6. Chaplin, M. F., (2006). *Water Structure and Behavior* (Vol. 1, pp. 1–30). http://www.lsbu.ac.uk/water/ (accessed on 3 March 2020).
7. Chen, X. D., (2009). Drying and dried food quality. In *Processing Effects on Safety and Quality of Foods* (Vol. 1, pp. 232–340). CRC Press.
8. Chieh, C., (2005). Fundamental characteristics of water. *Handbook of Food Science, Technology, and Engineering-4 Volume Set*, *1*, 20–50.
9. Coles, R., McDowell, D., and Kirwan, M. J., (2003). *Food Packaging Technology* (Vol. 5, pp. 1–30). CRC Press.
10. Coultate, T. P., (2009). *Food: The Chemistry of its Components* (Vol. 1, pp. 10–40). Royal Society of Chemistry.
11. Crosby, N. T., (1981). Food packaging materials. *Aspects of Analysis and Migration of Contaminants* (Vol. 1, pp. 5–20). Applied Science Publishers Ltd.
12. Cybulska, E. B., and Doe, P. E., (2006). Water and food quality. In *Chemical and Functional Properties of Food Components* (Vol. 1, pp. 39–70). CRC Press.
13. Da Pieve, S., Calligaris, S., Nicoli, M. C., and Marangoni, A. G., (2010). Shear nanostructuring of monoglyceride organogels. *Food Biophysics*, *5*(3), 211–217.
14. Dahl, L. W., and Andersen, H. C., (1983). A theory of the anomalous thermodynamic properties of liquid water. *The Journal of Chemical Physics*, *78*(4), 1980–1993.
15. DeMan, J. M., Finley, J. W., Hurst, W. J., and Lee, C. Y., (1999). *Principles of Food Chemistry* (Vol. 1, pp. 111–162). Gaithersburg,

MD: Aspen Publishers.

16. Edsall, J. T., and McKenzie, H. A., (1983). Water and proteins. II. The location and dynamics of water in protein systems and its relation to their stability and properties. *Advances in Biophysics*, *16*, 5–7.
17. Eisenberg, D., and Crothers, D., (1979). *Physical Chemistry with Applications to the Life Sciences* (No. 541.3 E3, pp. 5–20).
18. Eisenberg, D., and Kauzmann, W., (2005). *The Structure and Properties of Water* (Vol. 1, pp. 1–30). Oxford University Press on Demand.
19. Engberts, J. B. F. N., (2007). Structure and properties of water. *Organic Reactions in Water, 1*, 20–29.
20. Eubanks, L. P., (2006). *Chemistry in Context: Applying Chemistry to Society* (Vol. 1, pp. 1–30). Dubuque, IA: McGraw-Hill.
21. Fennema, O. R., Damodaran, S., and Parkin, K. L., (2017). Introduction to food chemistry. In *Fennema's Food Chemistry* (Vol. 3, pp. 1–16). CRC Press.
22. Fox, R. W., and McDonald, A. T., (1994). *Introduction to Fluid Mechanics* (Vol. 1, pp. 1–20), John WileyandSons. *Inc., New York.*
23. Franks, F., (2007). *Water: A Matrix of Life* (Vol. 1, pp. 1–20). Royal Society of Chemistry.
24. Gallier, J., Rivet, P., and De Certaines, J., (1987). 1H-and 2H-NMR study of bovine serum albumin solutions. *Biochimica et Biophysica Acta (BBA)-Protein Structure and Molecular Enzymology*, *915*(1), 1–18.
25. Gun'ko, V. M., and Turov, V. V., (1999). Structure of hydrogen bonds and 1H NMR spectra of water at the interface of oxides. *Langmuir*, *15*(19), 6405–6415.
26. Gun'ko, V. M., Turov, V. V., Bogatyrev, V. M., Zarko, V. I., Leboda, R., Goncharuk, E. V., and Chuiko, A. A., (2005). Unusual properties of water at hydrophilic/hydrophobic interfaces. *Advances in Colloid and Interface Science*, *118*(1–3), 125–172.
27. Hardy, J., Scher, J., and Banon, S., (2002). Water activity and hydration of dairy powders. *Le Lait*, *82*(4), 441–452.
28. Hatakeyama, H., and Hatakeyama, T., (1998). Interaction between water and hydrophilic polymers. *Thermochimica Acta*, *308*(1–2), 3–22.
29. Huffman, A. S., Afrassiabi, A., and Dong, L. C., (1986). Thermally reversible hydrogels: II. Delivery and selective removal of substances

from aqueous solutions. *Journal of Controlled Release, 4*(3), 213–222.

30. Hui, Y. H., Cross, N., Kristinsson, H. G., Lim, M. H., Nip, W. K., Siow, L. F., and Stanfield, P. S., (2006). Biochemistry of seafood processing. *Food Biochemistry and Food Processing, 1*, 351–378.
31. Isengard, H. D., (2001). Water content, one of the most important properties of food. *Food Control, 12*(7), 395–400.
32. Jenkins, S., and Morrison, I., (1999). Characterization of various phases of ice on the basis of the charge density. *The Journal of Physical Chemistry B, 103*(50), 11041–11049.
33. Jiang, S., and Lee, T., (2006). Freezing seafood and seafood products: Principles and applications. *Food Science and Technology* (Vol. 149, No. 1, p. 39). New York-Marcel Dekker.
34. Karel, M., and Yong, S., (1981). Autoxidation-initiated reactions in foods. In *Water Activity: Influences on Food Quality* (Vol. 1, pp. 511–529). Academic Press.
35. Kasaai, M. R., (2008). A review of several reported procedures to determine the degree of N-acetylation for chitin and chitosan using infrared spectroscopy. *Carbohydrate Polymers, 71*(4), 497–508.
36. Kasaai, M. R., (2010). Determination of the degree of N-acetylation for chitin and chitosan by various NMR spectroscopy techniques: A review. *Carbohydrate Polymers, 79*(4), 801–810.
37. Kireev, V. A., Peck, G., and Leib, F., (1972). *Physical Chemistry* (Vol. 1, pp. 1–20). Mir Publishers.
38. Kirillin, V. A., (1986). The present status and future introduction to the power industry of magnetohydrodynamic plants. *Thermal Engineering, 33*(2), 59–67.
39. Labuza, T. P., (1977). The properties of water in relationship to water binding in foods: A review. *Journal of Food Processing and Preservation, 1*(2), 167–190.
40. Le Bail, A., (2004). Freezing processes: physical aspects. *Food Science and Technology* (Vol. 1, pp. 1–12). New York-Marcel Dekker.
41. Lerner, K. L., and Lerner, B. W., (2006). *World of Forensic Science* (Vol. 1, pp. 1–20). Thomson/Gale.
42. Lewicki, P. P., (2004). Water as the determinant of food engineering properties. A review. *Journal of Food Engineering, 61*(4), 483–495.

43. Lewicki, P. P., (2009). Data and models of water activity. I: Solutions and liquid foods. In *Food Properties Handbook* (Vol. 1, pp. 48–81). CRC Press.
44. Lewicki, P. P., (2009). Data and models of water activity. II: Solid foods. In *Food Properties Handbook* (Vol. 1, pp. 82–167). CRC Press.
45. Li, J., and Ross, D. K., (1993). Evidence for two kinds of hydrogen bond in ice. *Nature, 365*(6444), 327.
46. Lide, D. R., (2004). *CRC Handbook of Chemistry and Physics* (Vol. 85, pp. 1–40). CRC press.
47. Liu, W. G., and De Yao, K., (2001). What causes the unfrozen water in polymers: Hydrogen bonds between water and polymer chains?. *Polymer*, *42*(8), 3943–3947.
48. Makkonen, L., (1997). Surface melting of ice. *The Journal of Physical Chemistry B*, *101*(32), 6196–6200.
49. Mathlouthi, M., (2013). *Food Packaging and Preservation* (Vol. 1, pp. 1–30). Springer Science and Business Media.
50. Rahman, M. S., and Sablani, S. S., (2009). Water activity measurement methods of foods. In *Food Properties Handbook* (Vol. 1, pp. 24–47). CRC Press.
51. Robinson, D. S., and Eskin, M., (2000). *Food Shelf Life Stability: Chemical, Biochemical, and Microbiological Changes* (Vol. 1, pp. 1–30). CRC Press.
52. Rockland, L. B., and Beuchat, L. R., (1987). Water activity: Theory and applications to food. In *IFT Basic Symposium Series (USA)* (Vol. 1, pp. 1–30). M. Dekker.
53. Roos, Y. H., (1993). Water activity and physical state effects on amorphous food stability. *Journal of Food Processing and Preservation*, *16*(6), 433–447.
54. Roos, Y. H., (2002). Importance of glass transition and water activity to spray drying and stability of dairy powders. *Le Lait*, *82*(4), 475–484.
55. Sammon, C., Deng, C., and Yarwood, J., (2003). Polymer-water interactions. Origin of perturbed infrared intensities of water in polymeric systems. *Polymer*, *44*(9), 2669–2677.
56. Slade, L., Levine, H., and Reid, D. S., (1991). Beyond water activity: recent advances based on an alternative approach to the assessment of food quality and safety. *Critical Reviews in Food Science and Nutrition*, *30*(2–3), 115–360.

57. Stanga, M., (2010). *Sanitation: Cleaning and Disinfection in the Food Industry* (Vol. 1, pp. 1–20). John Wiley and Sons.
58. Suresh, S. J., and Naik, V. M., (2000). Hydrogen bond thermodynamic properties of water from dielectric constant data. *The Journal of Chemical Physics, 113*(21), 9727–9732.
59. Van Oss, C. J., (2008). *The Properties of Water and Their Role in Colloidal and Biological Systems* (Vol. 16, pp. 1–20). Academic Press.
60. Welti, J., Bermúdez, D., Valdez, A., and Alzamora, S., (2004). *Principles of Freeze-Concentration and Freeze-Drying* (Vol. 2, pp. 13–24). New York.
61. Wiggins, P. M., (1971). Water structure as a determinant of ion distribution in living tissue. *Journal of Theoretical Biology, 32*(1), 131–146.
62. Zemansky, M. W., Abbott, M. M., and Van, N. H. C., (1975). *Basic Engineering Thermodynamics* (Vol. 1, pp. 1–30). McGraw-Hill Companies.

Chapter 5

Characterization of Proteins

CONTENTS

5.1. INTRODUCTION

Associated to chemical active components, biopharmaceutical active components are far more composite with regard to size, structure, sensitivity, and size to several degradation paths. Thus, particular analytical tools were created to classify these active components, which in many cases are therapeutic proteins (Yu, 2001). Amongst the therapeutic proteins, one portion of molecules, monoclonal antibodies, have a high perspective in a number of medical symptom areas (Srebalus Barnes and Lim, 2007; Kamerzell et al., 2011).

The biological action of bio-pharmaceuticals is described by the conformational and 3-D system of the molecule. Alterations of, for example, the protein secondary system might have a strong influence on its biological action, chemical along with toxicity and colloidal stability. Thus, this chapter describes some industrial applications and the latest developments in the area of protein analytical biophysics and biochemistry (Hawe et al., 2008; Philo, 2009).

Because biopharmaceuticals are formed through biotechnological procedures using living organisms, for example, mammalian cells, micro-heterogeneity of the created active component has to be considered (Spellman, 1990; Haydon et al., 2010). For issuance of the commercial products, control on quality is most important. A comprehensive report of the therapeutic protein product features is necessary. Diverse quality parameters have to be described to assure the stability, safety, and therapeutic action of the product (Kueltzo et al., 2003; Bobst et al., 2008; Pepys et al., 2012).

5.2. PROTEIN CHARACTERIZATION METHODS

Protein characterization comprises the utilization of experimental procedures that allow for the isolation and detection of protein and its cleansing, along with the classification of its function and structure. The success of new advanced, sensitive techniques and methods was the outcome of the latest advancements made in biotechnology, molecular medicine, molecular biology, biochemistry, and other associated sciences (Hamada, 1997; Nettleship et al., 2008). These were especially formed to allow the purification of peptides, proteins, amino acids, and metabolites. The scope of techniques utilized for protein purification can differ extensively ranging from simple one-step precipitation processes to large-scale authenticated production processes. To reach the chosen purity step more than is frequently necessary. The key for purification success is choosing the most suitable methods, to enhance and combine them in a logical manner to reduce required purification stages and to enhance yields. Most purification methods include some type of chromatography. Thus, chromatography has become a vital tool for protein purification (Baum et al., 1989; Sofia et al., 2001). The combination of diverse chromatography methods with changed selectivity permits for the design of dominant purification systems to refine any biomolecule (Löbler and Klämbt, 1985).

The growth of recombinant deoxyribonucleic acid (DNA) methods revolutionized the manufacturing of proteins in large numbers. Frequently recombinant proteins are formed such as this facilitates their succeeding chromatographic purification. This methodology has made purification simple but all challenges are not removed. Often impurities are still present and problems associated with structural integrity biological activity, solubility can still occur. Yet, in many purification systems, a protein might be purified first to a homogeneous product (Tillmann et al., 1989; Fekete et al., 2015).The scrutiny of the resultant protein products can be challenging and complex. Though, accurate classification of the last protein that was purified from natural sources or stated in different cell cultures or as a portion of a drug growth process is an important step. The complex systems and large size of proteins, along with the inherent nature of each protein, makes the classification of proteins intrinsically more complex than the classification of smaller molecules. Moreover, micro heterogeneities are usually observed even in extremely purified protein segments. Procedure- and product-related contaminants or impurities can make the purification procedure tough. Checking the last protein product for all potential impurities is compulsory. Every new purification procedure requires the execution of suitable specifications. Sometimes, to confirm reproducibility, even a simple analytical assessment, like quantification, can become very complex. During protein purification, expression, and functional in-vivo and in-vitro assess several analytical questions and control on quality to be addressed (Lenaz et al., 1968; Kamerzell et al., 2011).

5.3. PROTEIN CHARACTERIZATION APPROACHES

Proteins are composite molecular entities resultant from biological procedures. They vary from each other in their molecular structure, physicochemical properties, and size. These alterations permit for protein scrutiny and classification by identification and separation (Patterson and

Aebersold, 1995). In the production procedure, they are nearly always not fully homogeneous molecules and purified, but instead display a specific degree of structural heterogeneity. This means that the categorization of a protein always has three features (Ducret et al., 1998):

a- Identifying the major protein constituent;

b- Identifying the minor constituents; and

c- Quantifying the minor constituents (both product and process-related impurities).

Separation is usually done through electrophoresis where proteins are distinguished by mass or size and isoelectric focusing, where protein is detached by charge. These methods can be done in combination or independently mentioned to as 2D electrophoresis. Apparatus for separation contains organic stains, isoelectric focusing immobilized strips, electrophoresis boxes, 2D electrophoresis equipment protein standards and polyacrylamide gels (Szponarski et al., 2004; Glogauer et al., 2011). Identification is done through mass spectrometry (MS) where molecules are ionized to define their mass to charge ratios. MS apparatus includes LC/MS, CE/MS, GC/MS, time of flight systems and protein fractionators. Identification can be done through amino acid sequencing by crystal imaging, surface plasmon resonance, and Edman degradation. With various options available, laboratories usually work on multiple approaches for protein characterization and analysis (Kelly and Lansbury, 1994; Kueltzo et al., 2003). There are five major areas of protein characterization described below:

Protein quantification;

a. Glycosylation;

b. Protein modification;

c. impurity Characterization;

d Higher-order structure; and

e. Protein primary structure.

5.4. PROTEIN QUANTIFICATION

Exact protein quantitation is essential to understand the entire protein content in a formulated product or a sample and is necessary to study protein in different research topics. A wide range of different techniques has been formed to measure complex mixtures of proteins along with a single form of protein. How much protein is prevailing? It might look an easy question, but in actual protein, quantification is a tough task. Being a complex measurement, large mistakes are easily made and there are only a partial number of precise methods available (Cheng et al., 2008; Fortin et al., 2009; Ohtsuki et al., 2012).

Chemists and researchers had work a lot into developing and enhancing the techniques we use. Now, specially developed methods are available that provide high precision quantities and accuracy. Moreover, techniques can be cross-validated technique so that we don't have to put our word for it (Xu et al., 2008; Bergmann et al., 2012). Following techniques of protein quantification are widely utilized (Wang et al., 2007; Balogh et al., 2012):

a. Amino Acid Analysis (AAA), to calculate the amino acid constituents of a protein without utilizing an external standard. Determination;
b. Extinction Coefficient (A280), to evaluate protein concentration by quantifying Ultra-Violent absorbance at A280. This method helps to reveal comparability and consistency among batches. Note that protein absorption can also be described by MS and ELISA assays utilizing absolute quantification (AQUA). Protein quantification is the main part of ICH Q6B;
c. Extinction Coefficient (A280); and
d. Amino Acid Analysis (AAA).

5.4.1. AAA (Amino Acid Analysis)

Evaluate content with self-confidence-precise quantification of pure proteins and peptides is important for proteomics, systems biology, clinical chemistry, and biotechnology. Sometime during the assignment to create a biopharmaceutical drug or further research application, it is essential to find out accurate protein quantities (El-Ella et al., 1998). The reference technique to measure proteins and peptides is AAA The protein content must be calculated as precisely as possible as it is such an important parameter that will affect entirely other quantification outcomes got from other methods. Even minor deviance can lead to inappropriate results and finally, to an inappropriate product. (Kotha and Brahmachary, 2004; Glew et al., 2008).

This is where PPS (Protagen Protein Services) could support the clients. Standard AAA containing hydrolysis, is a great tool for evaluating the absorption of a sample or the accurate protein content without utilizing an exterior standard. But AAA is an even more influential tool when utilized with a time-resolved hydrolysis technique, which will provide the exact amino acid composition of a protein. This method is normally used for the classification of a reference standard (Hanemaaijer et al., 1998).

5.4.2. Technical Information

Technical particulars of AAA are illustrated below:

a. AAA (Amino acid analysis) can be applied in several ways to disclose different features of the protein. Here are certain instances. In combination with ultra-violent-absorbance quantities, it can be utilized to directly conclude the extinction quantity of a protein.
b. If hydrolysis is skipped, it could be utilized to calculate free amino acids, for instance in cell culture media.
c. It can measure unusual amino acids such as Hydroxyproline, -lysine, and Norleucine.
d. To calculate concentration founded on mass, it could be combined with (MALDI-TOF) matrix-assisted laser desorption/ionization-time of flight mass spectrometer, to disclose the complete mass of the protein.

5.5. PROTEIN MODIFICATION

One of the significant parts of the protein characterization is to recognize those proteins that are post-translational revised and their modification sites and to describe the function of the alterations (Baeuerle and Huttner, 1987; Klahr et al., 1995). Nearly all proteins, whether purified or produced as recombinant proteins or insulated from the natural sources, would take some degree of revised amino acids.Characterizing product associated contaminations is the main task. One has to investigate them with much accuracy and sensitivity as possible. Such amendments are part of the physiological system that we are forming (Glomb and Monnier, 1995).

However one is interested to witness the minimum possible number of procedure-related alterations. The last might happen during the sample handling procedure, for instance in proteins that have been kept in a stability study for a specific period. They are unintended and a cause of structural heterogeneity in a protein (Gilmore et al., 2006).

All such alterations are known as post-translational alterations and whereas some of them are wanted, others are unwanted.

One can select from the following kinds of analysis:

a. Deamidation;

bn. Glycosylation;

c. Phosphorylation;

d. Oxidation;

e. Acetylation;

f N- and C-terminal truncation; and

g Pyroglutamate formation.

Describing post-translational protein amendments is the main requirement of ICH Q6B.

a Deamidation;

b Other Modifications; and

c. Oxidation.

5.5.1. Deamidation

The distress with small alterations-The distress with tiny alterations in a complicated molecule is tough in trying to discover them. This is the contest when proteins especially those comprising the amino acids glutamine or asparagine had experienced deamidation. It outcomes in the reduction or addition of as tiny as one Dalton of the molecular weight (MW) to complete the protein system and it's very tough to pick up. Then there is the problem of trying to work out if the alteration is actually because of deamidation or simply due to changes in isotopic distribution, which is usual and happens naturally in the peptides. Investigating deamidation is the main requirement of ICH Q6B.

5.5.2. Technical Information

Deamidation is usually observed as a post-translational alteration of the amino acid asparagine, which converts into a blend of aspartate and iso-aspartate through a succinimide intermediate. The amount of alteration is impacted by factors like the interaction and presence of surrounding amino acids, temperature buffers, and pH. Deamidation of glutamine also happens, but significantly less frequently.One is interested in discovering whether the protein had to go through deamidation as it leads to alteration in (pI) isoelectric point and thus the existence of charge heterogeneity. One can consistently observe pI through IEX-HPLC and 1D-IEF.

5.6. GLYCOSYLATION

Proteins that had to go through glycosylation are extremely complex. Many glycans might be attached to several amino acids, and every glycan will display unique and occasionally unexpected branching (Seela and Bourgeois, 1989; Trachtman et al., 1994).

Though glycosylation analysis is a relatively new topic of research however we can apply a variety of methods, depending on the situations, to get a complete outline of the glycans in proteins. They contain MS of high resolution, Electrophoretic, and HPLC methods. (Sato et al., 2007; Kohda et al., 2009).

MS is a significant method due to its high sensitivity and selectivity. It also gives information rapidly. The MS plan depends on the stage of structural data which is required. A usual strategy is shown below.

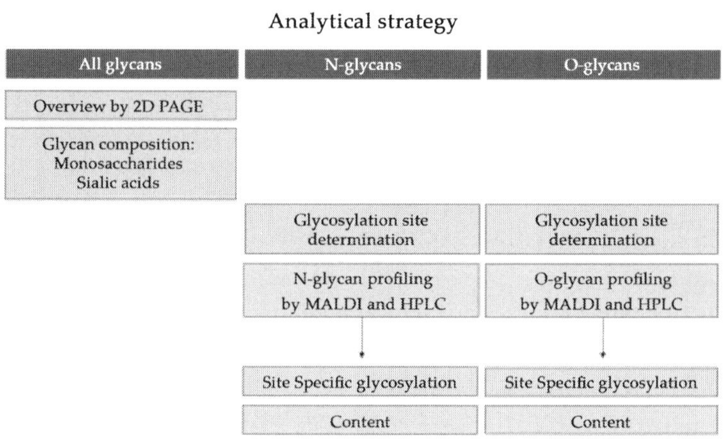

Figure 5.1: Analytical approach for glycosylation.

One can select from the following kinds of analysis:

a. N-linked glycosylation, to observe the attachment of N-linked glycans at the conclusion of enzymes;
b. O-linked glycosylation, to examine the more complicated addition of O-linked glycans, which found later in the processing of protein;
c. Sialic acid content, to describe the quantity and type of sialic acid exist at any level of drug development, info which is useful in

further glycosylation examination;

d. Monosaccharide content, to evaluate the quantity and composition of O- and N-linked glycosylation; and

e. Glycosylation site, to work on sites precisely where glycans are linked to the proteins, because usually not all accessible sites will be full.

Protein glycosylation is a post-translational alteration that changes a protein's function or structure. It is a very normal and biologically related alteration that might directly involve in the creation of diseases. It also plays a significant role in changing the pharmacokinetics by improving solubility or protecting from proteases and steadying the protein conformation.

The key challenge in the examination of protein glycosylation is its structural complications, which rises from a combination of reasons like as:

a. N-linked glycosylation;
b. `O-linked glycosylation;
c. The complex, branched system of glycans;
d. Sialic acid content;
e. Multiple glycan systems attached to each site;
f. Multiple glycosylation sites per protein;
g. Glycosylation site; and
h. Monosaccharide analysis.

5.6.1. N-linked Glycosylation

When related to O-linked glycosylation, the investigation of N-linked glycosylation is just straightforward, as scientists had been exploring it for a longer period. The most precise, most economical and fastest method is through utilizing mass spectrometers.

For instance, one might like to describe the heterogeneity and structure of glycosylation for a whole protein or every potential glycosylation location. To describe the system of N-glycans we'll utilize MS (LC-ESI or MALDI) after permethylation or labeling. But to describe the antennary of glycans, or to measure glycans, one will utilize HPAEC-PAD. Being a usual HPLC technique, it has the benefit that one can simply confirm the assess for relief testing. The Normal Phase HPLC (NP-HPLC) matches this situation by giving a highly sensitive, additional, and high-resolution scrutiny of the samples.

Locations of glycan attachment regulate either by peptide MALDI-ISD or mapping.

Additional information about a protein's glycosylation can be joined with what one recognizes about the N-linked glycans. For instance, one can define the amount and composition of glycans attached to a protein through sialic acid analysis and monosaccharide analysis. The system heterogeneity of a glycoprotein can be envisaged utilizing intact mass or 2D PAGE.

5.6.2. Technical Information

N-linked glycans are a normal feature of glycoproteins like FSH, EPO, and antibodies. They have revealed to be significant in various ways. They have an impact on protein folding along with other factors, like immune reactions, stability, cell-to-cell interactions, and protein solubility.

N-glycosylation enhances the structural heterogeneity of proteins when a variety of structurally similar-however not identical-glycans are attached to every glycosylation location.

Glycosylation might be directly comprised in the creation of diseases, but also plays a significant part in changing the pharmacokinetics through improving solubility, protecting from proteases and stabilizing protein conformation.

The N-glycosylation sketch of a recombinant protein can differ largely relying on the organism selected for expression.

5.7. PROTEIN PRIMARY STRUCTURE

A protein is additional than the total of its amino acids-When viewing at a protein in drug development, the first phase we most probably want to do is an examination of its covalent bonds and amino acids or primary system. Following areas are discovered to determine the structure of a protein (Sundelin et al., 1985; Zimniak et al., 1992):

a. Intact mass;

b. N/C terminal sequencing;

c. De-novo sequencing;

d. Peptide Mapping; and

e Disulfide linkage.

Not fair a protein identity check-peptide mapping together is a regulatory obligation under ICH Q6B and an influential tool for accepting how a protein might be improved to outfit the purpose. It gives a complete set of information to update the drug development procedure (Persico et al., 1986).

Peptide mapping is usually started to examine the protein primary system after cleavage of the protein into proteolytic peptides. The influence of peptide mapping found in a large number of location precise molecular features that can be identified.

When utilizing one digestion enzyme (for instance, Trypsin), peptide mapping is usually done for protein recognition. The analysis is done utilizing LC-ESI-MS/MS or MALDI-MS/MS for instance during protein empathy after electrophoresis, like a 2D PAGE or 1D.

When utilizing multiple enzymes, peptide mapping is placed for the validation of a complete amino acid order, for instance when authorizing the amino acid order of a biosimilar and relating it with the initiator molecule.

Relying on the experimental system, peptide mapping can also be utilized to decide the C- and N-terminus of a protein. This is seldom essential information for the customers, for instance in the case of monoclonal antibodies, where the truncation level of the C-terminal Lysine can be checked.

Peptides and Proteins are fragmented and ionized in the mass spectrometer. The resulting MS/MS spectra are utilized to order individual amino acids in the sequence of their appearance in a peptide or protein.

In combination with added sample preparation, peptide mapping can also be utilized to determine:

a. Glycosylation sites and structure;
b. Disulfide linkages;
c. The degree of oxidation or deamidation for every affected amino acid; and
d. N-glycosylation sites.

If the protein classification is not found in a protein record, it can be assumed by de-novo sequencing.

Up to 100% sequence exposure confirms the amino acid series. The identification is founded on LC-ESI-MS/MS figures, which authorize the sequencing of peptides with great sureness. Overlying peptide sequences endorse the arrangement of amino acids in the sequence.

5.8. IMPURITY CHARACTERIZATION

Despite purification correspondingly polishing processes in the protein preparation, it is not probable to form proteins without host cell proteins. In addition to these procedure-related contaminations, there are also product interrelated contaminations to be caused by protein alterations. Protein contaminations may produce undesired side effects (Horak et al., 2010; Levy et al., 2014).

One can quantify and identify all the proteins in the blend utilizing MS. Through performing out a bioassay, one can also observe if the contaminations are influencing the effectiveness. Lastly, the undesirable proteins are removed one phase at a time by concentrating on the differences in their features (Xie et al., 2009).

5.8.1. Applications in Pharma-Biotech

a. Characterization and optimization of Down-Stream and Up-Stream Processing (DSPandUSP) steps in the situation of protease burden;
b. Comparability of protein-based or protein-containing drugs;
c. Assessment of the existence of proteases in drug product affecting stability problems; and
d. Characterization of protein-based or protein-containing drugs.

5.8.2. Application in Diagnostics

1. Quality control of complicated buffers or the last immunoassay utilized as a diagnostic kit; and
2. Quality control of protein formed raw materials like IgG molecules or Serum Albumin.

5.9. HIGHER-ORDER STRUCTURE

Developing the creation of biopharmaceutical could be encouraging and concrete knowledge of the system and conformation of the product is necessary. Correct HOS (higher-order structure) is acute to ensure proper stability, activity, and functionality of a biopharmaceutical product. A well-formed panel of procedures for HOS classification is a necessary component of a whole product classification program (Gregory and Dahlberg, 1999; Wei et al., 2014).

Different creations might alter the tertiary and secondary system of a biopharmaceutical, which in response can distress protein activity. PPS gives analytical techniques to analyze the stability and structure of the biopharmaceutical in its creation.

a. Analytical ultracentrifugation (AUC);
b. Circular Dichroism (CD); and
c. Differential Scanning Calorimetry (DSC).

AUC (analytical ultra-centrifugation) splits protein types directly in solution, without the utilization of a stationary stage like as in (SEC) size-exclusion chromatography. The sedimentation frequency of the molecule(s) is convinced by the centrifugal force and is observed constantly by fluorescence, or interferometry and UV absorbance to develop a size distribution summary of the species present inside the test sample. (SV-AUC) Sedimentation velocity AUC produces sedimentation coefficient principles and reports the comparative percentages of multimer, monomer, and aggregate species. Information on the hydrodynamic shape and MW is also obtained (Nishi et al., 1994).

5.10. FLUORESCENCE

Fluorescence uses the natural inherent fluorescence of Tryptophan (Trp) and Tyrosine (Tyr) remains and gives information on the domestic environments around these remains. Fluorescence is a brilliant relative tool and is paired with FT-IR and CD (Stork et al., 1968; Rosowsky et al., 1986; Malby et al., 2001).

The beneath image shows the comparative inherent fluorescence of Tyr and Trp remains for a product in its preparation buffer IR (InfraRed) spectra gives quantitative and qualitative information on the secondary system of proteins like α β sheets, β turns, helices, and disordered structures.

The most instructive IR bands for the protein examination are amide I (1620 to 1700 cm^{-1}), amide II (1520 to 1580 cm^{-1}) and amide III (1220 to 1350 cm^{-1}). Amide I is the most concentrated absorption band in proteins and contains the stretching shaking of the C=O (70 to 85% and C-N groups (10 to 20%).

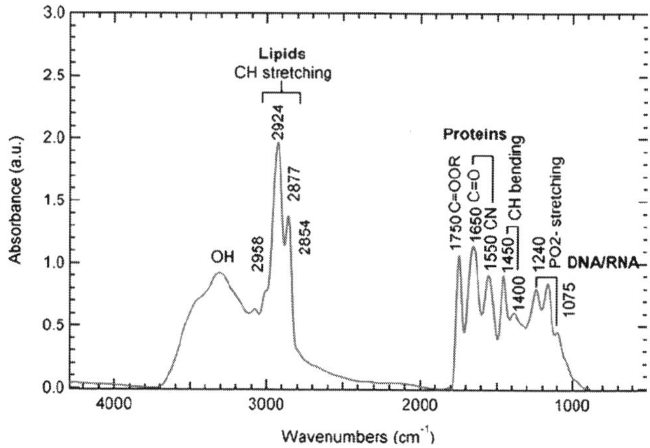

Figure 5.2: FTIR spectra of proteins.

Amide II is regulated through in-plane N-H twisting (40 to 60%), C-C (10%), C-N (18 to 40%) stretching vibrations. FT-IR gives an orthogonal valuation of a secondary system to Far UV CD analysis. It is frequently considered more beneficial than CD for products with a high range of β sheets and α helices because, distinct Far UV CD, FT-IR doesn't express a disproportionately high reply to α helix. The profiles got and fitting the data can be utilized to evaluate the comparability of the secondary system of different formulations or batches and amongst biosimilar and originator samples. Overlay of the FT-IR spectra at 2200 to 1000 cm^{-1}.

REFERENCES

1. Baeuerle, P. A., and Huttner, W. B., (1987). Tyrosine Sulfation is a trans-Golgi-specific protein modification. *The Journal of Cell Biology*, *105*(6), 2655–2664.
2. Balogh, L. M., Kimoto, E., Chupka, J., Zhang, H., and Lai, Y., (2012). Membrane protein quantification by peptide-based mass spectrometry approaches: Studies on the organic anion-transporting polypeptide family. *J. Proteomics Bioinform*, *4*, 3.
3. Baum, J., Dobson, C. M., Evans, P. A., and Hanley, C., (1989). Characterization of a partly folded protein by NMR methods: Studies on the molten globule state of guinea pig. Alpha-lactalbumin. *Biochemistry*, *28*(1), 7–13.
4. Bergmann, U., Ahrends, R., Neumann, B., Scheler, C., and Linscheid, M. W., (2012). Application of metal-coded affinity tags (MeCAT): Absolute protein quantification with top-down and bottom-up workflows by metal-coded tagging. *Analytical Chemistry*, *84*(12), 5268–5275.
5. Bobst, C. E., Abzalimov, R. R., Houde, D., Kloczewiak, M., Mhatre, R., Berkowitz, S. A., and Kaltashov, I. A., (2008). Detection and characterization of altered conformations of protein pharmaceuticals using complementary mass spectrometry-based approaches. *Analytical Chemistry*, *80*(19), 7473–7481.
6. Cheng, F. Y., Blackburn, K., Lin, Y. M., Goshe, M. B., and Williamson, J. D., (2008). Absolute protein quantification by LC/MSE for global analysis of salicylic acid-induced plant protein secretion responses. *Journal of Proteome Research*, *8*(1), 82–93.
7. Ducret, A., Oostveen, I. V., Eng, J. K., Yates, III. J. R., and Aebersold, R., (1998). High throughput protein characterization by automated reverse-phase chromatography/electrospray tandem mass spectrometry. *Protein Science*, *7*(3), 706–719.
8. El-Ella, W. A., Baky, A. A., Aly, M. E., and Fox, P. F., (1988). Effect of ripening temperatures on proteolysis and lipolysis in the outer and inner regions of ras-type cheese made by various salting methods. *Food Chemistry*, *28*(1), 1–16.
9. Fekete, S., Guillarme, D., Sandra, P., and Sandra, K., (2015). Chromatographic, electrophoretic, and mass spectrometric methods for the analytical characterization of protein biopharmaceuticals.

Analytical Chemistry, *88*(1), 480–507.

10. Fortin, T., Salvador, A., Charrier, J. P., Lenz, C., Bettsworth, F., Lacoux, X., and Lemoine, J., (2009). Multiple reaction monitoring cubed for protein quantification at the low nanogram/milliliter level in nondepleted human serum. *Analytical Chemistry*, *81*(22), 9343–9352.
11. Gilmore, J. M., Scheck, R. A., Esser-Kahn, A. P., Joshi, N. S., and Francis, M. B., (2006). N-terminal protein modification through a biomimetic transamination reaction. *Angewandte Chemie. International Edition*, *45*(32), 5307–5311.
12. Glew, R. S., Chuang, L. T., Roberts, J. L., and Glew, R. H., (2008). Amino acid, fatty acid and mineral content of black finger millet (Eleusine coracana) cultivated on the Jos Plateau of Nigeria. *Food*, *2*(2), 115–118.
13. Glogauer, A., Martini, V. P., Faoro, H., Couto, G. H., Müller-Santos, M., Monteiro, R. A., and Krieger, N., (2011). Identification and characterization of a new true lipase isolated through metagenomic approach. *Microbial Cell Factories*, *10*(1), 54.
14. Glomb, M. A., and Monnier, V. M., (1995). Mechanism of protein modification by glyoxal and glycolaldehyde, reactive intermediates of the Maillard reaction. *Journal of Biological Chemistry*, *270*(17), 10017–10026.
15. Gregory, S. T., and Dahlberg, A. E., (1999). Erythromycin resistance mutations in ribosomal proteins L22 and L4 perturb the higher order structure of 23 S ribosomal RNA. *Journal of Molecular Biology*, *289*(4), 827–834.
16. Hamada, J. S., (1997). Characterization of protein fractions of rice bran to devise effective methods of protein solubilization. *Cereal Chemistry*, *74*(5), 662–668.
17. Hanemaaijer, R., Janssen, A., De Kok, A., and Veeger, C., (1988). The dihydrolipoyl transacetylase component of the pyruvate dehydrogenase complex from *Azotobacter vinelandii:* Molecular cloning and sequence analysis. *European Journal of Biochemistry*, *174*(4), 593–599.
18. Hawe, A., Sutter, M., and Jiskoot, W., (2008). Extrinsic fluorescent dyes as tools for protein characterization. *Pharmaceutical Research*, *25*(7), 1487–1499.
19. Haydon, D. J., Bennett, J. M., Brown, D., Collins, I., Galbraith, G., Lancett, P., and Nayal, N., (2010). Creating an antibacterial with in

vivo efficacy: synthesis and characterization of potent inhibitors of the bacterial cell division protein FtsZ with improved pharmaceutical properties. *Journal of Medicinal Chemistry*, *53*(10), 3927–3936.

20. Horak, J., Ronacher, A., and Lindner, W., (2010). Quantification of immunoglobulin G and characterization of process related impurities using coupled protein A and size exclusion high performance liquid chromatography. *Journal of Chromatography A*, *1217*(31), 5092–5102.
21. Kamerzell, T. J., Esfandiary, R., Joshi, S. B., Middaugh, C. R., and Volkin, D. B., (2011). Protein-excipient interactions: Mechanisms and biophysical characterization applied to protein formulation development. *Advanced Drug Delivery Reviews*, *63*(13), 1118–1159.
22. Kelly, J. W., and Lansbury, P. T., (1994). A chemical approach to elucidate tin mechanism of transthyretin and β-protein amyloid fibril formation. *Amyloid*, *1*(3), 186–205.
23. Klahr, S., Breyer, J. A., Beck, G. J., Dennis, V. W., Hartman, J. A., Roth, D., and Yamamoto, M. E., (1995). Dietary protein restriction, blood pressure control, and the progression of polycystic kidney disease. Modification of Diet in Renal Disease Study Group. *Journal of the American Society of Nephrology*, *5*(12), 2037–2047.
24. Kohda, Y., Kanematsu, M., Kono, T., Terasaki, F., and Tanaka, T., (2009). Protein O-glycosylation induces collagen expression and contributes to diabetic cardiomyopathy in rat cardiac fibroblasts. *Journal of pharmacological sciences*, 0912040306–0912040306.
25. Kotha, S., and Brahmachary, E., (2004). Synthesis and reactions of silicon containing cyclic α-amino acid derivatives. *Journal of Organometallic Chemistry*, *689*(1), 158–163.
26. Kueltzo, L. A., Ersoy, B., Ralston, J. P., and Middaugh, C. R., (2003). Derivative absorbance spectroscopy and protein phase diagrams as tools for comprehensive protein characterization: A bGCSF case study. *Journal of Pharmaceutical Sciences*, *92*(9), 1805–1820.
27. Kueltzo, L. A., Ersoy, B., Ralston, J. P., and Middaugh, C. R., (2003). Derivative absorbance spectroscopy and protein phase diagrams as tools for comprehensive protein characterization: A bGCSF case study. *Journal of Pharmaceutical Sciences*, *92*(9), 1805–1820.
28. Lenaz, G., Haard, N. F., Lauwers, A., Allmann, D. W., and Green, D. E., (1968). Mitochondrial structural protein I: Methods of preparation and purification: Characterization by gel electrophoresis. *Archives of Biochemistry and Biophysics*, *126*(3), 746–752.

29. Levy, N. E., Valente, K. N., Choe, L. H., Lee, K. H., and Lenhoff, A. M., (2014). Identification and characterization of host cell protein product-associated impurities in monoclonal antibody bioprocessing. *Biotechnology and Bioengineering*, *111*(5), 904–912.

30. Löbler, M., and Klämbt, D., (1985). Auxin-binding protein from coleoptile membranes of corn (Zea mays L.). I. Purification by immunological methods and characterization. *Journal of Biological Chemistry*, *260*(17), 9848–9853.

31. Malby, S., Pickering, R., Saha, S., Smallridge, R., Linse, S., and Downing, A. K., (2001). The first epidermal growth factor-like domain of the low-density lipoprotein receptor contains a noncanonical calcium binding site. *Biochemistry*, *40*(8), 2555–2563.

32. Nettleship, J. E., Brown, J., Groves, M. R., and Geerlof, A., (2008). Methods for protein characterization by mass spectrometry, thermal shift (ThermoFluor) assay, and multiangle or static light scattering. In *Structural Proteomics* (Vol. 1, pp. 299–318). Humana Press.

33. Nishi, K., Yoshida, M., Fujiwara, D., Nishikawa, M., Horinouchi, S., and Beppu, T., (1994). Leptomycin B targets a regulatory cascade of crm1, a fission yeast nuclear protein, involved in control of higher order chromosome structure and gene expression. *Journal of Biological Chemistry*, *269*(9), 6320–6324.

34. Ohtsuki, S., Schaefer, O., Kawakami, H., Inoue, T., Liehner, S., Saito, A., and Terasaki, T., (2012). Simultaneous absolute protein quantification of transporters, cytochromes P450, and UDP-glucuronosyltransferases as a novel approach for the characterization of individual human liver: comparison with mRNA levels and activities. *Drug metabolism and Disposition*, *40*(1), 83–92.

35. Patterson, S. D., and Aebersold, R., (1995). Mass spectrometric approaches for the identification of gel-separated proteins. *Electrophoresis*, *16*(1), 1791–1814.

36. Pepys, M. B., Gallimore, J. R., Lloyd, J., Li, Z., Graham, D., Taylor, G. W., and Millar, D. J., (2012). Isolation and characterization of pharmaceutical grade human pentraxins, serum amyloid P component and C-reactive protein, for clinical use. *Journal of Immunological Methods*, *384*(1–2), 92–102.

37. Persico, M. G., Viglietto, G., Martini, G., Toniolo, D., Paonessa, G., Moscatelli, C., and D'urso, M., (1986). Isolation of human glucose-6-pbosphate debydrogenase (G6PD) cDNA clones: Primary structure of

the protein and unusual 5'non-coding region. *Nucleic Acids Research*, *14*(6), 2511–2522.

38. Philo, J. S., (2009). A critical review of methods for size characterization of non-particulate protein aggregates. *Current Pharmaceutical Biotechnology*, *10*(4), 359–372.
39. Rosowsky, A., Wright, J. E., Cucchi, C. A., Boeheim, K., and Frei, III. E., (1986). Transport of a fluorescent antifolate by methotrexate-sensitive and methotrexate-resistant human leukemic lymphoblasts. *Biochemical Pharmacology*, *35*(2), 356–360.
40. Sato, S., Naito, Y., and Aoki, K., (2007). Scandium cation-exchanged montmorillonite catalyzed direct C-glycosylation of a 1,3-diketone, dime done, with unprotected sugars in aqueous solution. *Carbohydrate Research*, *342*(7), 913–918.
41. Schweitzer, B., Predki, P., and Snyder, M., (2003). Microarrays to characterize protein interactions on a whole-proteome scale. *Proteomics*, *3*(11), 2190–2199.
42. Seela, F., and Bourgeois, W., (1989). Stereoselective glycosylation of nitrobenzimidazole anions: Synthesis of 1,3-dideaza-2′-deoxyadenosine and related 2′-deoxyribofuranosides. *Synthesis*, *1989*(12), 912–918.
43. Sofia, H. J., Chen, G., Hetzler, B. G., Reyes-Spindola, J. F., and Miller, N. E., (2001). Radical SAM, a novel protein super family linking unresolved steps in familiar biosynthetic pathways with radical mechanisms: Functional characterization using new analysis and information visualization methods. *Nucleic Acids Research*, *29*(5), 1097–1106.
44. Spellman, M. W., (1990). Carbohydrate characterization of recombinant glycoproteins of pharmaceutical interest. *Analytical Chemistry*, *62*(17), 1714–1722.
45. Srebalus, B. C. A., and Lim, A., (2007). Applications of mass spectrometry for the structural characterization of recombinant protein pharmaceuticals. *Mass Spectrometry Reviews*, *26*(3), 370–388.
46. Stork, G., Kretchmer, R. A., and Schlessinger, R. H., (1968). The stero specific total synthesis of dl-lycopodine. *Journal of the American Chemical Society*, *90*(6), 1647–1648.
47. Sundelin, J., Das, S. R., Eriksson, U., Rask, L., and Peterson, P. A., (1985). The primary structure of bovine cellular retinoic acid-binding protein. *Journal of Biological Chemistry*, *260*(10), 6494–6499.

48. Szponarski, W., Sommerer, N., Boyer, J. C., Rossignol, M., and Gibrat, R., (2004). Large-scale characterization of integral proteins from Arabidopsis vacuolar membrane by two-dimensional liquid chromatography. *Proteomics*, *4*(2), 397–406.
49. Tillmann, U., Viola, G., Kayser, B., Siemeister, G., Hesse, T., Palme, K., and Klämbt, D., (1989). cDNA clones of the auxin-binding protein from corn coleoptiles (Zea mays L.): isolation and characterization by immunological methods. *The EMBO Journal*, *8*(9), 2463–2467.
50. Trachtman, H., Futterweit, S., Prenner, J., and Hanon, S., (1994). Antioxidants reverse the antiproliferative effect of high glucose and advanced glycosylation end products in cultured rat mesangial cells. *Biochemical and Biophysical Research Communications*, *199*(1), 346–352.
51. Wang, L., Xu, G., Shi, Z., Jiang, W., and Jin, W., (2007). Quantification of protein-based on single-molecule counting by total internal reflection fluorescence microscopy with adsorption equilibrium. *Analytica Chimica Acta*, *590*(1), 104–109.
52. Wei, H., Mo, J., Tao, L., Russell, R. J., Tymiak, A. A., Chen, G., and Engen, J. R., (2014). Hydrogen/deuterium exchange mass spectrometry for probing higher-order structure of protein therapeutics: Methodology and applications. *Drug Discovery Today*, *19*(1), 95–102.
53. Xie, H., Gilar, M., and Gebler, J. C., (2009). Characterization of protein impurities and site-specific modifications using peptide mapping with liquid chromatography and data-independent acquisition mass spectrometry. *Analytical Chemistry*, *81*(14), 5699–5708.
54. Xu, M., Yang, L., and Wang, Q., (2008). Quantification of selenium-tagged proteins in human plasma using species-unspecific isotope dilution ICP-DRC-q MS coupled on-line with anion exchange chromatography. *Journal of Analytical Atomic Spectrometry*, *23*(11), 1545–1549.
55. Yu, L., (2001). Amorphous pharmaceutical solids: Preparation, characterization, and stabilization. *Advanced Drug Delivery Reviews*, *48*(1), 27–42.
56. Zimniak, P., Eckles, M. A., Saxena, M., and Awasthi, Y. C., (1992). A subgroup of class α glutathione S-transferases Cloning of cDNA for mouse lung glutathione S-transferase GST 5.7. *FEBS Letters*, *313*(2), 173–176.

Chapter 6

Advanced Analysis of Carbohydrates in Foods

CONTENTS

6.1. INTRODUCTION

The most varied and abundant class of organic compounds are carbohydrates present in nature. In chemical nature, they are made up of hydrogen, carbon (C), and oxygen in the ratio of $C_{n:}H_{2n}:O_{n}$. According to their chemical, the food carbohydrates comprising a wide range of macromolecules can be classified into three main groups high molecular weight (MW) polysaccharides, intermediate MW oligosaccharides, and low MW disaccharides. They can also be categorized as complex or simple carbohydrates. Complex carbohydrates are monosaccharides such as fiber (polysaccharides) and starches while simple carbohydrates are made up of monosaccharides and disaccharides.

Figure 6.1: Foods with healthy carbohydrates.

Source: https://bluehousejournal.blogspot.com/2016/09/eat-health-part-ii-carbohydrates-energy.html.

In the human diet, the major source of energy is carbohydrates with consumptions ranging from 40 to 80% of total energy needs (Muir et al., 2009). For all body functions, carbohydrates make up for the main source of energy, especially brain functions, and are essential for the metabolism of other nutrients. Control of blood glucose, serum cholesterol, and insulin metabolism, gastric, and satiety emptying, influencing gastrointestinal processes such as fermentation and laxation influencing colonic microflora are other significant effects of carbohydrates on human physiology (Muir et al., 2009).

The major sources of carbohydrates are vegetables, potatoes, cereals, fruits, legumes, flour, and rice. Hence, the consumption of naturally occurring sugars is part of a healthy diet. Polysaccharides, sucrose, and monosaccharides are existent in all vegetables (Hounsome et al., 2008). Carbohydrates are produced by all green plants in the body are either stored in the form of glycogen or absorbed immediately. They can also be synthesized in the body from the glycerol component of fats and amino acids. Mainly to enhance the sensorial quality of food sugars can be added to foods during preparation or processing (Murphy et al., 2003).Food carbohydrates can be divided in two class according to nutritionists: readily utilized, available, and metabolize, including mono-, di,- or polysaccharides such as lactose, glucose, starch, and dextrin's, fructose, and *b*, carbohydrates not directly used but instead broken down by symbiotic bacteria producing fatty acids therefore not providing the host with carbohydrates or unavailable (Ninonuevo et al., 2005, 2006). Many complex polysaccharides such as pectin's, cellulose, and β-glucans and structural polysaccharides of plant cell walls are included in the second class (Noe et al., 1999; Cui, 2005). In other words, unavailable carbohydrates are those that not hydrolyzed by endogenous human enzymes (many oligosaccharides, sugar alcohols, and nonstarch polysaccharides) and available carbohydrates ones are hydrolyzed by enzymes of the human gastrointestinal system, but unavailable ones in the large intestine can be fermented by microorganisms to variable extents and then absorbed (Hounsome et al., 2008).Sugar polyols (mannitol and sorbitol), sugars (fructose, glucose, lactose, sucrose, and maltose), polysaccharides (non-starch and starch polysaccharides), oligosaccharides (FOS, fructooligosaccharides, and GOS, galactooligosaccharides) have been defined as the main classes of carbohydrates appropriate for human nutrition. The fermentable-short-chain carbohydrates like oligo-, di-monosaccharides, and polyols that can be rarely absorbed by the small intestine are known as FODMAPs and they deliver diverse effects on gastrointestinal health. FODMAPs can be established in a varied variety of foods (Okatch et al., 2003; Muir et al., 2009).

6.2. FOOD CARBOHYDRATES

Everyone can find detailed data regarding the chemistry of food carbohydrates (Cui, 2005). In the following section, the major features of the food carbohydrates appropriate for human nutrition are delivered together with the significance of developing informative and powerful methods for their adequate analysis.

6.2.1. Simple Carbohydrates

Disaccharides and monosaccharides are two simple kinds of carbohydrates. Monosaccharides can be absorbed straight into the bloodstream and require no digestion. The body can synthesize all the monosaccharides (Hounsome et al., 2008). Monosaccharide sugars are the basic carbohydrates, of which galactose, fructose, and glucose are the most significant nutritionally. Figure 6.2 shows the fructose structure, it is a major constituent of the honey, fruit juices, corn syrup and fruits (Park et al., 1993), and is also one of the major FODMAPs of the Western diet (Gibson et al., 2007).

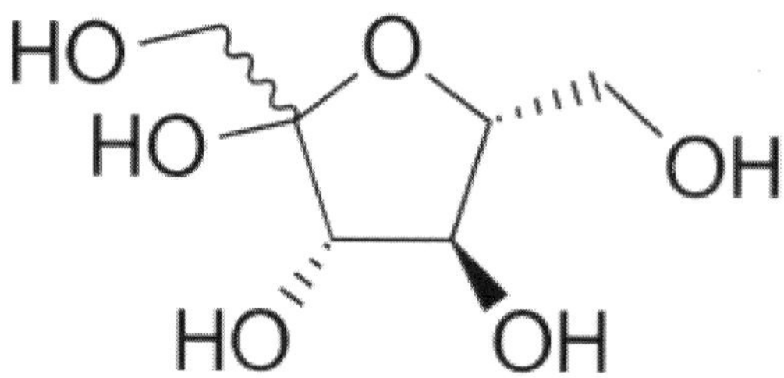

Figure 6.2: Fructose structure.

Source: https://en.wikipedia.org/wiki/Fructose.

Two monosaccharides combine to make disaccharides. The important disaccharides are lactose, a dimer of galactose and glucose; sucrose, a dimer of fructose and glucose; and maltose, a dimer of two glucose units. Lactose has also been categorized as FODMAP (Muir et al., 2009).

6.2.2. Sugar Polyols

Sugar polyols, also known as sugar alcohols, are acquired by the reduction of the equivalent ketoses and aldehydes. Sugar polyols can be used as low-calorie sweeteners since they have produced much interest as food additives. Two sweeteners very regularly used in the food industry are shown in Figure 6.3, such as sorbitol and mannitol structures (Peris-Tortajada, 2000).

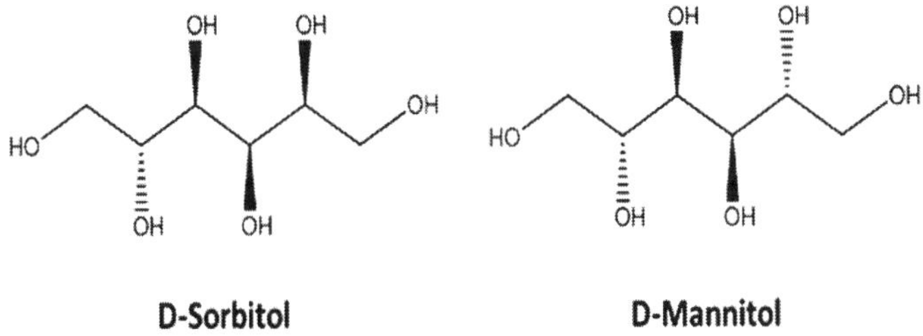

Figure 6.3: Chemical structures of D-sorbitol and D-mannitol.

Source: https://www.nature.com/articles/s41598-018-27450-w.

Though a very limited data have been published so far on levels of polyols in foods, it is identified that in various vegetables mannitol is existent such as parsley, celery, onion, carrot, asparagus, and endive (Hounsome et al., 2008) while sorbitol is also used as a sweetener and is existent in many fruits. It has been suggested to be used as indicators of the adulteration or authenticity of foods by some authors (Martínez-Montero et al., 2004; Ruiz-Matute et al., 2007a).

6.2.3. Oligosaccharides

Various oligosaccharides occur in foods, consisting of 3–5 monosaccharide units and in general, they are non-digestible. The definition of oligosaccharides can also be written as those carbohydrates produced by 2 to 10 basic sugar molecules. Galactooligosaccharides (GOS, stachyose, raffinose), fructooligosaccharides (FOS, as e.g., kestose, nystose), mannan oligosaccharides (MOS) are considered prebiotic and FODMAPs carbohydrates (Figure 6.4). Non-starch polysaccharides and FOS play a significant role are the dietary fibers (Hounsome et al., 2008; Miur et al., 2009).

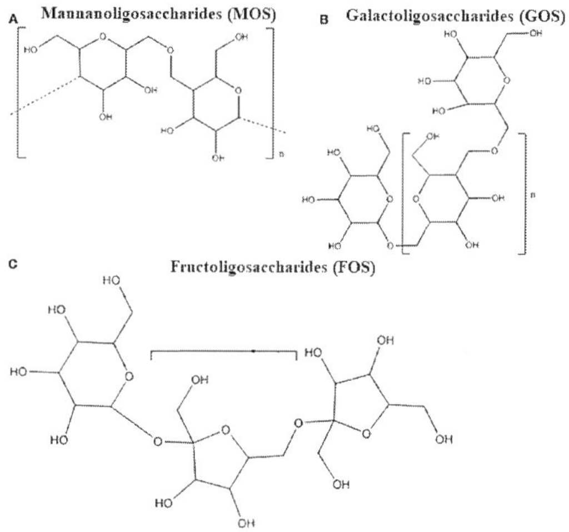

Figure 6.4: Structural formulae of (A) Mannanoligosaccharides (B) Galactoligosaccharides (C) FOS.

Source: https://www.researchgate.net/publication/327039175_A_Review_of_Prebiotics_Against_Salmonella_in_Poultry_Current_and_Future_Potential_for_Microbiome_Research_Applications/figures?lo=1andutm_source=googleandutm_medium=organic.

Jerusalem artichokes, garlic, and onions have high levels of fructans (Muir et al., 2007). The highest concentration of FOS has been found in the Jerusalem artichoke in cultured plants. Since the '90s in Japan FOS have been employed in dietary supplements and they are now fetching gradually popular for their prebiotic effects in Western cultures. In the large intestine, for microflora, FOS serves as a substrate, growing the overall gastrointestinal tract health. Furthermore, numerous studies have found that in both humans and animals gut the polysaccharide and FOS inulin stimulate calcium absorption. In lower gut the intestinal microflora can ferment FOS, producing the gases and in the decrease of pH that rises calcium bioavailability and solubility.

6.2.4. Polysaccharides

Polysaccharides are given a general term called Glycans in which by *O*-glycosidic linkages, large numbers of monosaccharides are naturally joined. Polysaccharides are considered condensation polymers formed by glycosidic linkage from the glycosyl moiety hemiketal or hemiacetal and a hydroxyl group acting as an aglycone or acceptor molecule of another sugar unit. Polysaccharides may be branched or linear (Figure 6.5). There are two main kinds of polysaccharides: heteropolysaccharides and homo-polysaccharides. A hetero-polysaccharide has two or more types of monosaccharides, whereas, homo-polysaccharide is composed of one kind of monosaccharide repeating in the chain. Unbranched polysaccharides comprise only alpha 1,4 linkages while some of these branched polysaccharides are connected to a molecule through alpha 1,4 and to one more one via alpha 1,6 glycosidic bonds. These macromolecules affect both nutrition and food quality and hence have a large physiological interest.

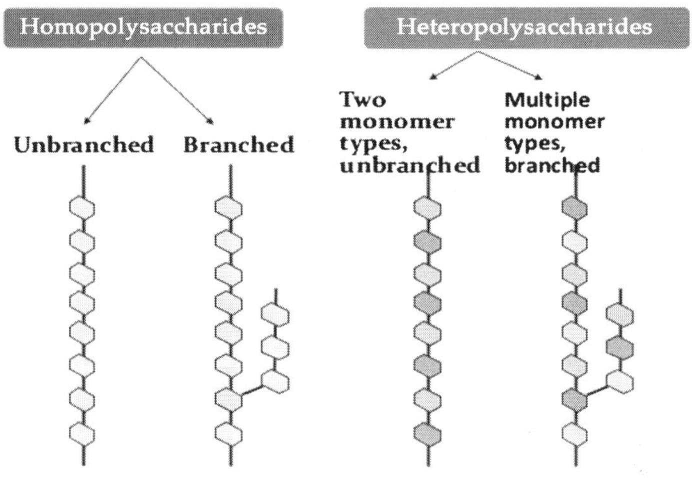

Figure 6.5: Chemical structures of different polysaccharide variants.

Source: https://www.slideshare.net/annuarunpatel/biosynthesis-of-polysaccharides.

6.3. SAMPLE PREPARATION FOR CARBOHYDRATES ANALYSIS

Parting from the rest of the main components is a crucial step when examining carbohydrates in food complex matrices, such as proteins and lipids that can somehow delay in their accurate quantification and determination. Once separated, carbohydrates can be either subjected to some additional treatments (such as derivatization and/or hydrolysis) or analyzed directly to favor their succeeding examination.

6.3.1. Extraction and Fractionation

Generally, some kind of clean-up or extraction step is used for the separation of carbohydrates from other food components prior to their analysis. However, in some conditions, the sample attained is still too complicated to be simply analyzed and a further fractionation stage is required to acquire a specific carbohydrate fraction. In this segment, the techniques used for fractionation and extraction of carbohydrates are described.

6.3.1.1. Filtration

Normally, a simple dilution and filtration step could be used for the preparation of the sample, when examining the carbohydrate fraction of a beverage. For example, this has been the situation in wines (La Torre et al., 2008), milk, and whey (Mullin at al. 1997). Even filtration and ultrafiltration have been used to concentrate carbohydrates based on the use of membrane technologies (Kamada et al., 2002). However, when the quantity of diverse class of carbohydrates is very dissimilar or the original matrix is too complex, more sophisticated procedures are required for the sample preparation. It is worth noting that as a clean-up, a filtration step can be used before the use of any other procedure mentioned below (Khandurina and Guttman, 2005).

6.3.1.2. Traditional Extraction Methods (LLE, SPE)

Currently, the so-called traditional extraction methods have been extensively applied such as solid-phase extraction (SPE) and liquid-liquid extraction (LLE) for the action of carbohydrates from food matrices before analysis. For instance, for the extraction of carbohydrates from beer (Lehtonen at al. 1994) wine (Castellari et al., 2000), honey (Weston et al., 1999), SPE has been used among others. According to the nature of the sample to be examined, the nature of the solvent employed for elution as well as the solid column packing is selected. C_{18} cartridges are perhaps the most employed (Megherbi et al., 2008), though others such as porous graphitic C (Arias et al., 2003), C_8 (Kitahara and Copeland, 2004), or even cartridges of the basis of strong cation exchange (Castellari et al., 2000) are also used. In fractionating complex samples, SPE has shown its ability: for example, celite columns and active C were employed to fractionate oligosaccharides and monosaccharides from honey by the use of various water/ethanol ratios to get a selective elution (Weston et al., 1999; Lion et al., 2006). While the outcomes that can be acquired using these methods might be agreeable, some significant deficiencies characteristic to their experimental principles, mostly connected to the excessive quantities of organic solvents desired, have brought about a rise in the use of other extraction methods (Klampfl Buchberger, 2001).

6.3.1.3. Supercritical Fluid Extraction (SFE)

The basis of SFE is the use of fluids above their critical temperature and pressure. Supercritical fluids obtain properties characteristics between gases (similar diffusivity) and liquids (similar solubility), under these conditions. Thanks to the moderate critical pressure and temperature of CO_2 which is most commonly used fluid, along with other significant advantages such as by change the operating pressure, the easiness to adjust the selectivity of the extraction, the fulfillment of an extract free of solvent and the likelihood to get fractions by carrying out dissimilar depressurization steps because CO_2 is discharged as a gas when pressure is reduced under ambient conditions (Kolhed Larlberg, 2005; Louw et al., 2008). Though, by its low polarity, its use in carbohydrate examination has been partially hampered, which makes the solubilization of these compounds difficult. Various quantities of the organic solvents used as modifiers are employed, to overcome the difficulties. To fractionate several carbohydrates, SFE has been used many times that could be employed in the food industry as prebiotics (Montañés at al. 2007; Montañés et al., 2006). Mainly the selectivity was affected by the co-solvent composition as the authors demonstrated in these works, while other parameters such as the amount of polar cosolvent, pressure, and the temperature influenced the carbohydrates recovery. It is anticipated that following similar approaches other applications of this method to the fractionation and extraction carbohydrates can be established.

6.3.1.4. Pressurized Liquid Extraction (PLE)

The basis of PLE is the usage of solvents at high pressures and temperatures during the whole extraction process high enough to retain the liquid state. The extraction processes ingest fewer quantities of solvents, produce higher yields and the extraction processes are faster compared to the more traditional extraction methods. PLE is being extensively used to extract sample components from natural products and foods, due to these advantages (Mendiola at al. 2007). Various solvents such as organic solvents, water, and mixtures can be employed. Few works have been published, as for carbohydrates fractionation relating to the selective lactulose extraction from the mixture by the use of 30:70 water/ethanol mixture for 30 minutes at 40°C (Ruiz-Matute et al., 2007b). Moreover, enriched fractions of di- and trisaccharides were acquired by using PLE from honey combined with the use of activated charcoal (Ruiz-Matute et al., 2008). In this situation, honey was packed into a PLE extraction cell after being absorbed onto activated charcoal that was given in to two different extraction cycles. The second one was the use of 50:50 water/ethanol for 10 minutes and the first one was using 99:1 water/ethanol for 5 minutes. From other sources, carbohydrates fractionation can also be done by the use of this PLE procedure.

6.3.1.5. Field-Flow Fractionation (FFF)

The basis of this technique is the separation of sample components being pushed by a laminar stream inside an open channel in which a perpendicular (magnetic, electric, gravitational or thermal) field is applied. The fractionation or separation occurs according to the changes in the movement of the components as a consequence of the applied field. This method is often employed to fractionate large polysaccharides, for carbohydrates analysis, that could not be then fractionated mainly starch, pullulan, cellulose, among others.

6.3.1.6. Chromatography-Based Methods

To separate and fractionate carbohydrates, various methods based on the use of chromatographic procedures have been used. Hence, the most popular method is size exclusion chromatography (SEC) for the measurement of molecular mass distributions of carbohydrates (Eremeeva, 2003). To separate and fractionate oligo- and polysaccharides, SEC has been extensively used due to these features, including the analysis of carbohydrates in rice (Ward et al., 2006), the examination of prebiotic galactosaccharides (Hernandez et al., 2009), coffee (Arya at al. 2007), wheat (Hartmann at al. 2005) or sorghum (Aboubacar et al., 2000).

Ion Exchange Chromatography (IEC) is another method based on chromatography extensively used to fractionate carbohydrates. IEC is typically suitable for mono- and oligosaccharides fractionation. Depending on the specific application, both cation and anion exchange resins can be applied. The normal methodology is to apply IEC as a methodical step rather than for fractionation and, hence, IEC will be debated in more detail below (Herrero et al., 2009).

Simulated moving bed chromatography (SMBC) is the last method included in this group. In the sugar industry, this method has been extensively used for large-scale palatinosetrehaulose fractionation. From the constant counter-current parting of a sample, two streams of products (extract and raffinate) are acquired eluted from a series of various adsorbents. The basis of the absorbents can be SEC or ligand exchange chromatography. To fractionate lactose from a complex mixture, this method has been used for human milk oligosaccharides (Geisser at al. 2005).

6.3.2. Chemical Treatments

In the preparation of the sample for the food carbohydrates, a various chemical process is generally used to enable their subsequent examination.

6.3.2.1. Hydrolysis

Generally, before the chromatic examination, to obtain fractions of low molecular oligo- or monosaccharides, a hydrolysis step is carried out of large carbohydrates. In the existence of a strong acid under heating forming a polysaccharide by the glycosidic bonds joining the various monosaccharides can be cleaved. Hydrochloric acid, trifluoroacetic acid and even sulphuric acid are the most commonly used acids. Temperature, acid concentration, time, and other treatment conditions, have to be adjusted to attain total hydrolysis of the polysaccharides present in the sample. Parallel to this, these conditions can generate sample degradation because these are too strong, the released monosaccharides are disposed to degradation since it is well acknowledged that during this procedure (Tkac et al., 2009). As an instance, the use of 2 M HCl for 90 min at 100°C is the typical condition applied to the procedure of the wheat flour. Before analysis polysaccharides can also acquiesce to incomplete hydrolysis. Partial degradations can be attained at organized conditions, thanks to the differential vulnerability to hydrolysis of the various glycosidic linkages to yield mixtures of various oligo- and monosaccharides whose examination could be valuable to measure the polysaccharide structure. Recently, to reduce the procedure time to hydrolyzate carbohydrates microwave irradiation has been used as a tool (Corsaro et al., 2004; Cortacero-Ramirez et al., 2004).

6.3.2.2. Derivatization

It is not easy to analyze carbohydrates by various separation techniques, given their intrinsic properties because of their very similar polarity, low volatility or lack of chromophore. Hence, to analyze carbohydrates by GC, they have to be derivatized to increase their volatility. Monosaccharides can be examined by as alditol acetates by GC attained by acetylation and reduction, while oligosaccharides can be analyzed as trifluoroacetyl or trimethylsilyl ethers by GC (Corzo-Martinez et al., 2010).

There are two techniques which require care about sensitivity, i.e., HPLC or CE, because UV light below 195 nm can only be absorbed by the vast majority of intact carbohydrates, or can be detected using the refractive index (RI) or amperometric detectors. Hence, fluorescent or UV-absorbing labels can be sued to raise the sensitivity of their detection, in several derivatization procedures. Being the more common process pre-column derivatization, a reaction may occur during or even before or the analysis (Lamari et al., 2003; Molnar-Perl, 2000). The usage of reductive amination is the most common technique applied. Condensation of a carbonyl group is the basis of these processes are to produce a Schiff base with primary amines that are reduced later on to yield an N-substituted glycosyl amine. The most commonly employed reagents are 2-aminopyridine of trisulfonates. Also taking into account the fact that most of the carbohydrates are hydrophilic and neutral compounds, derivatization yields electrical charge and polarity changes along with sensitivity enhancement that can make the separation easier by HPLC or CE. In literature, a detailed review can be found on this topic (Lamari et al., 2003; Molnar-Perl, 2000; Suzuki et al., 2003).

6.4. ADVANCED ANALYSIS OF FOOD CARBOHYDRATES

The main tools employed to analyze the carbohydrates in food separation techniques, spectroscopic techniques and they are multiple, though some other approaches are also significant in this area based on the new glycomics platforms or the use of sensors. Based on the relevance and novelty of papers these have been selected for citation, trying to give an up to date overview to the reader about the analysis of carbohydrates.

6.4.1. Spectroscopic Techniques

For quantitative and qualitative analysis of food ingredients and food, spectroscopic methods have been used for years, mainly using infrared (IR), visible (Vis), atomic absorption, fluorescence, electron spin resonance (ESR), mass spectrometry (MS), Raman spectroscopy (RS), nuclear magnetic resonance (NMR), ultraviolet (UV), and atomic emission (Ibáñez et al., 2003; Momenbeik et al., 2006).

Nowadays, for the advancement of various foodomics methods, there is an obvious trend to use high-resolution MS and NMR instruments (Cifuentes, 2009), which includes metabolomics. Metabolomics tries to identify and detect compounds i.e., amino acids, flavonoids, vitamins, carbohydrates, glucosinolates, hormones, and phenolics which are necessary for the animal and plant development and growth, defense, and stress adaptation and also for the measurement of the sensorial and nutritional quality of food. Hence, MS and NMR methods give inspiring possibilities in food carbohydrates examination and various structural studies which involve the application of both MS and NMR for the characterization and identification of food components.

6.4.1.1. Nuclear Magnetic Resonance (NMR)

For an accurate characterization of both the natural carbohydrates and newly designed the use of NMR is becoming critical as the structure of the carbohydrate will define its function. The information related to the purity, safety, and structure of carbohydrate samples is being provided by the application of NMR. For the identification of food carbohydrates and related compounds, the NMR has widely been employed. This analytical method has shown its feasibility for the study of simple and complex carbohydrates either combined with food matrices or isolated (Montañes et al., 2006, 2007; Hounsome et al., 2008). Because of their various applications in the food industry (health-promoting, sweetener, etc.) Now, oligosaccharides naturally comprising the food matrices and those newly designed are of great importance. As a result of the verified influenced of nutrition in the health and the demand of food consumers, there is more interest in the exploration of low cost, bioactive, natural compounds are called functional foods and to be used as health-promoting food ingredients. The most useful oligosaccharides are FOS, though, for the properties of non-insulin-type fructans, there is a lack of information. Recently, the use of the H^1 NMR along with enzymatic hydrolysis indorsed identify fructans in crushing fractions of two wheat cultivars (Haslam et al., 2006; Haska et al., 2008). Arabinoxylan, an essential indigestible component in muesli grains with useful health effects, was proved as a key component of the dietary fiber of all fractions. The corresponding characterization of the fructans, like the determination of MW, involved the usage of other methods for the structural study like high-performance anion interchange chromatography with MALDI-TOF-MS and pulse amperometric detection (HPAECPAD) (Haska et al., 2008).

In the preparation of specialty foods for people with specific health concerns, synthetic carbohydrates can also be used. The presence of lactose and glucose, for example, maybe dangerous for lactose intolerants and diabetics. As an example, new FOS made from sucrose by the use of fungal fructose transferase have been known by using LC-MS and NMR spectroscopy (Mabel et al., 2008). The quantification and detection of disaccharides and monosaccharides in the FOS preparations to be used as non-nutrients sweeteners were also accepted in this work. Furthermore, the spectral data found through the application of these two analytical methods shown the nonexistence in the carbohydrate preparation of any toxic microbial metabolites certificating its safety as a non-nutrient sweetener with probable profits to diabetics (Mabel et al., 2008).

NMR has been also effectively used for structural characterization of trisaccharides found by enzymatic transglycosylation of lactulose. Authors use either 2D (gCOSY, ROESY, TOCSY, gHMBC, and gHSQC) 1D and (^{1}H and ^{13}C TOCSY) NMR methods. The structure of the two trisaccharides attained by the enzymatic action of lactulose was explained and C assignment and full proton were carried out. Authors have suggested the enzymatic transgalactosylation as a new method for prebiotic synthesis based on the structural data derived from NMR analysis, (Martínez-Villaguenga et al., 2008). NMR has been also used as an instrument for searching economically feasible sources for mass manufacture of pinitol, a sugar alcohol that keeps hypoglycaemic effect. The study intricates the analysis of 115 normal and food materials. Based on the NMR outcomes, the authors proposed carob pod and soy whey as economically feasible sources of pinitol (Kim et al., 2003, 2009a, b).

By application of NMR characterization of very compound molecules comprising carbohydrates alike, coffee melanoidins have been also assumed. Melanoidins give to the flavor and color of coffee and are supposed to have physiological action. The use of the 2D-NMR method allowed us to get novel information supporting the idea that coffee melanoidins are centered on condensed polyphenols. Data is in opposition to those earlier published representing that they can be made by low-molecular-weight chromophores bound to proteins or polysaccharides (Gniechwitz et al., 2008). It is assumed that the spectrometric methods, and mainly NMR, are very vital for the analysis of carbohydrates related to proteins and other physiological related food components (Takahashi et al., 1995; Taga et al., 2001).

Other compound carbohydrates such as polysaccharides have been identified by ^{13}C and 1H NMR. For example, the main polysaccharides of hazelnut cell walls, have been determined by NMR (Dourado et al., 2003); polysaccharides composition of medicinal mushroom and an immature tomato has been also just characterized by using different NMR tryouts (Chandra et al., 2009). Several papers associated with carbohydrate metabolomic studies in animal foods and vegetables have been also described (Karakach et al., 2009; Hounsome et al., 2008). In the complex food carbohydrates field, it is estimated an increase in the number of applications of NMR.

6.4.1.2. Mass Spectrometry (MS)

MS is an influential analytical technique used to classify chemical compounds built on their different mass-charge ratio (m/z). In this section, just the use of direct MS analysis to food carbohydrates is explained. The usage of the mass detectors combined with various isolation methods will be discussed in the following sections (Godshall et al., 2001; Hsu and Mahal, 2009).

Before entering the MS analyzer the chemical compound has to be ionized; chemical ionization (CI), electron impact ionization (EI), plasma desorption ionization (PDI), electrospray ionization (ESI), and matrix laser desorption ionization (MALDI) are the most commonly used ionization methods for the analysis of organic compounds like food carbohydrates. High-quality reviews giving huge and thorough information associated with the state of the art of MS analysis of complex, simple, and conjugated carbohydrates have been published (Zaia, 2004; Harvey, 2008).

Because of their multiple blends, low volatility and complex nature, the structural analysis of carbohydrates has been a tough task to be assumed; so, the functional implications of the carbohydrates are poorly identified. At present, MS is starting to be reflected vital for fast characterization of functions and structures of these macromolecules below the so-called field of glycomics (Ibañez Cifuentes, 2003; Zhai et al., 2007).

Using mass spectroscopy, it is critical to select the right ionization technique and matrix for gaining good structural results (Lastovickova et al., 2009; Cmelik et al., 2004). For example, for the detection of larger oligosaccharides, MALDI-TOF-MS is thought to be more sensitive, while for the structural studies ESI-MS seems to be more useful (Cmelik et al., 2004). Isolated carbohydrates can be analyzed by the use of MS techniques important in food chemistry. Careri et al. revised the applications of MS techniques examination of oligosaccharides in several foods in 2002. Recently, for the characterization of oligosaccharides, MALDI-TOF-MS has been employed in some vegetable samples (red onion, Jerusalem artichoke and glucose syrup from potatoes); the degree of polymerization of the oligosaccharides in each food was recognized, once the spectrometric circumstances were improved. For the detection of oligosaccharides, MALDI-TOF-MS is more sensitive by the authors than the chromatographic technique use for a similar purpose (Stikarovska et al., 2004; Jager et al., 2007), underlying the significance of MS for food carbohydrate classification. According to the incredible influence of MS on food chemistry and glycomics, it is anticipated a larger number of MS applications for the examination of food carbohydrates.

6.4.2. Gas Chromatography (GC)

Carbohydrates' determination in foods have been extensively done by GC. GC has been mostly applied considering the high diversity of carbohydrates in terms of functionality, size, and structure and their low volatility to study tri-di- and monosaccharides composition of foods. Though, using a temperature programming getting 360°C, oligosaccharides with polymerization degree up to 7 have been effectively measured (Montilla et al., 2006; Sharon, 2008). Nowadays, for the GC analysis of carbohydrates most used derivatives are ethers Trimethylsilyl oximes (Rojas-Escudero et al., 2004). However, to detect closely related species acylated derivatives are favored, such as sugar alcohols (mannitol, sorbitol, galactitol) or aldohexoses (mannose, glucose, and galactose) which only vary in one or more hydroxyl groups.

Normally, dimethylpolysiloxane stationary phases are employed for the separation of sugars by GC with a given percentage of polar phenyl groups, with maximum programmed degrees getting characteristically 300–320°C. Flame ionization detector (FID) is the most commonly used detection method (Tomiya et al., 1998; Torto et al., 2000). Detector use of mass spectrometers increased more recently, giving various benefits over the traditional FIDs, largely, in the identification of carbohydrates has been made possible to a higher degree, thanks to the reproducible fragmentation patterns and the determination of the molecular mass. To identify the carbohydrates, a normal approach comprises the use of both, a GC-MS method and also to carry out the quantification of the use of GC-FID. Minor carbohydrates in carrots were quantified and identified using this approach (Rovio et al., 2007; Soria et al., 2009) and coffee (Ruiz-Matute et al., 2007a). It was possible to identify sedoheptulose in the carrot chemical composition and the trimethylsilyl oximes of scylloinositol for the first time, thanks to the MS detector employed, along with the other known slight carbohydrates myo-inositol and mannitol.

Honey, whose composition depends on its floral origin is a carbohydrate-rich food. To evaluate the specific composition of unifloral honey and their origin, the analysis of the honey contained carbohydrates, as well as their relationships, have been used (de la Fuente et al., 2007). In this case, statistical analyses have verified its effectiveness for correct sample assignment such as principal component analysis (PCA) (Terrab et al., 2001; Cotte et al., 2004).

To identify unknown carbohydrates, the application of GC-MS has been also revealed to be valuable. In different analytical columns, the disaccharides composition of numerous honey was measured. The elution behavior of an unidentified disaccharide was studied by the use of two different columns. This was compared with commercial standards. The existence of inulobiose could be confirmed by combining all this information and its fragmentation in the MS (Ruiz-Matute et al., 2007c).

GC has also been used to study the composition of carbohydrates of fruits and fruit-related products. Together with flavonoids, carboxylic acids and sugar alcohols, tri-, di-, and monosaccharides from fruits have been studied by GC using MS (Fuzfai et al., 2004) and FID (Bartolozzi et al., 1997) detectors. In several citrus juices, various flavonoids together with di- and monosaccharides, amino, and carboxylic acid and sugar alcohol could be directly quantified and recognized after sample hydrolysis (Fuzfai et al., 2007).

Under heat treatment, this disaccharide is produced by lactose isomerization, which is another example of the determination of the lactulose in commercial milk. Depending on intensity and type of heat treatment, numerous studies displayed its various concentrations; thus this can be used as an example of the heat treatment procedures in the UHT and sterilized commercial milk. Various methods have been explained to determine lactulose, among them, the GC examination of their equivalent trimethylsilyl ethers separated on 50% dimethylsiloxane /50% diphenyl capillary column. On different treated milk it was possible to compute lactulose, using this protocol (pasteurized, UHT, powder, and condensed milk, sterilized) (Montilla et al., 2005). UHT milk has a significantly higher amount of lactulose as demonstrated by authors than those determined in pasteurized samples, hence, it showed that this compound could be used as a marker for heat treatment processes in commercial samples (Montilla et al., 2005). For detecting food adulterations, carbohydrate determination can also be used in this regard (Cotte et al., 2003).

During heat treatments, maltulose can also be produced as a result of isomerization from maltose and has been recommended as a potential marker for the process of heat treatment. To assess heat treatment, the ratio of maltose/maltulose determined after the GC analysis was recommended as potential during the storage and manufacture of enteral and infant formulas (García-Baños et al., 2002; Morales et al., 2004, 2006). Quality markers in sugar-rich foods have been the application of di-fructose anhydrides such as coffee and honey (Montilla et al., 2006b). During caramelization reactions that occur under, condensation of two fructose molecules produces these compounds. Various samples treated under different heat treatment conditions were analyzed by the GC analysis of these compounds (Montilla et al., 2006b).In end, the suitable application of this method is inadequate by the essential sample preparation steps and its use to relatively small carbohydrates, although GC offers excellent outcomes. In fact, during all these processes, there is a fair chance of losses in the sample, mainly due to hydrolysis not finished or/and derivatization reactions. Other methods, such as HPLC or CE have been established trying to overcome these restrictions (Garcia-Cañas and Cifuentes, 2008).

6.4.3. High-Performance Liquid Chromatography (HPLC)

For carbohydrates analysis, the main analytical technique used is HPLC. HPLC can offer some extra benefits, compared to the other analytical methods. For example, HPLC is faster, although less sensitive than GC, does not essentially need sample derivatization, permits the examination of larger carbohydrates and it is more multipurpose thanks to the detectors that can be employed and diverse separation mechanisms (Morales-Cid et al., 2007).

6.4.3.1. HPLC Separation Mechanisms

Depending on the sample to be examined, the instrumental obtainability, and the precise carbohydrates of interest, the separation mechanism can be selected. Partition (reversed or normal phase) chromatography, ion-exchange chromatography and SEC are the most used HPLC modes for carbohydrate analysis.

Using cationic or anionic exchange resins ion-exchange chromatography can be used. Especially, in carbohydrates analysis, high-performance anion-exchange chromatography (HPAEC) has found great achievement (Cataldi et al., 2000). Quaternary ammonium resins are commonly used to carry out the separation to allow carbohydrates ionization under highly alkaline conditions, though nowadays different kinds of commercial columns are accessible. The mobile phase influences the selectivity as well as the total analysis time of the separation, hence it must be carefully selected. Oligosaccharides can be examined by HPAEC besides, di-, and monosaccharides. According to their polymerization degree, separation of oligosaccharides can be done by various methods developed so far. For example, maintaining 10% 1 M Na+ hydroxide constant during the entire analysis using a gradient from 1 M Na+ acetate to deionized water, oligosaccharides from corn syrups and honey samples can be studied (Morales et al., 2008). (Morales et al., 2008). Various malto-oligosaccharides were sufficiently separated (from polymerization degree 1 to 16) using this technique. After sample hydrolysis poly- and oligosaccharides can also be examined by HPAEC to make simpler the complex carbohydrates profile to a simpler di and monosaccharides composition.

To separate oligo- and monosaccharides, cationic metal ions, i.e., Ag^+, or Ca^{2+} have been also used from various food matrices. Formation of weak complexes between the carbohydrates hydroxyl groups and the metal ions are the basis of the retention mechanism. This specific separation technique permits the whole separation of monosaccharide enantiomers, also called ligand exchange chromatography, such as glucose, galactose, or mannose. Also, sugar alcohol can be appropriately separated. Though, this method has numerous shortcomings: metals can be eluted from the resins so columns are not very robust when analyzing complex samples and due to the very weak interaction with the metal ions the resolution of oligosaccharides is frequently poor.SEC is another separation approach extensively used for food carbohydrates. In the sample preparation section, this technique has already been mentioned since it is also employed to fractionate carbohydrates. Furthermore, it has been allowed its use as a high-performance liquid chromatography-based procedure by the technological growths permitting the engineering analytical columns.

The specific structure and pore size of resin determine the permeation of substances which is the basis of SEC. Polysaccharides, cannot penetrate inside the resins, being the three-dimensional network of large molecules and thus are eluted first. According to reducing molecular size, the rest of the molecules will elute. Hence, monosaccharides will interact more with the resins if they are smaller in size and thus will take a longer time to elute. For example, this method has been used for the chemical classification of arabinoxylans from cereals (Hartmann et al., 2005). Arabinoxylans are connected with various technological features in bread making as well as with different organoleptic parameters of bread. To measure the rye and wheat arabinoxylans mass distribution, SEC was used together with a UV detector during the bread-making procedure.

Partition chromatography is the latter separation process that has to be stated, either inverted or stage partition. Based on the use of the HPLC process, the count of requests is increasing in the area of column technology due to infamous enhancements and expansions that have been made. In this way, many application is being found by the original stationary phases made of spherical silica particles bonded with trifunctional aminopropyl silane in carbohydrate analysis. This unusual situation of normal-stage separations is called hydrophilic interaction liquid chromatography (HILIC). The primary functionalized amino-bonded silica gel columns that were utilized to discrete mono- and disaccharides have revealed to be less resilient than these functionalized amino columns. In several fruits and vegetables for the analysis of fermentable polyols and oligo-, di-, and monosaccharides, including glucose, fructans, FOS, galactooligosaccharides, and different sugar polyols. This column was eluted with acetonitrile/water (75:25, v/v) (Muir et al., 2009). For the examination of oligosaccharides from chickpeas (Xiaoli et al., 2008; Yagi and Kato, 2009), based on secondary-tertiary amines, with another column type, the same moveable phase was employed. Though the outcomes attained are not, in general, as acceptable as with the amino columns but still, more standard C_{18} RP-HPLC columns have been also utilized. Besides, in this latter case, the derivatization of the carbohydrates may be essential (Yagi et al., 2005).

6.4.3.2. HPLC Detectors

The detection mode is the parameter to consider since its choice will largely affect the sample preparation to be performed when examining carbohydrates by HPLC. To allow their discovery and/or to raise sensitivity, it will be essential to succumb the carbohydrates to a derivatization step if a photometric detection is successful to be used (UV-V fluorescence). There are two main groups of Carbohydrates derivatization reactions; the first is based on the outline of chromophores reacting with hydroxyl groups and the second includes the reaction with amines at the carbonyl group. 1-phenyl-3-, ethyl-5pyrazolone is one of the most used derivatizing reagents to discuss UV absorbance. Without the usual necessities of an acidic catalyst, which evades degradation and unwanted isomerization this compound can react with carbohydrates under mild conditions, providing a sensitive UV detection at 245 nm (Zhang et al., 2003). Other reagents that are also used are phenyl isocyanate (detection at 240 nm) or p-nitrobenzol chloride (absorbance at 260 nm).

Though pre-column derivatization reactions are used, post-column derivatization has also been defined. In the existence of reducing carbohydrates in alkaline medium and high-temperature benzamidine was used as a derivatizing agent to form a fluorescent complex. This reaction was applied by HPLC in different food-related materials for the concurrent determination of mono- and oligosaccharides (Kakita et al., 2002). The 288 nm as excitation wavelength and 470 nm for emission were being used for fluorescence detection, resulting in LOD of 2.59 and 1.78 pmol for maltohexaose and D-glucose respectively. The same process was carried out to find reducing sugars as well as phosphorylated sugars in chicken meat. In this way, tetrazolium blue was the selected reagent, which conferred absorbance at 550 nm (Aliani et al., 2002).

Other discovery systems have been typically used in food carbohydrate analysis to evade difficult derivatization steps and it reflects difficult photometric detection. RI detection is extensively been used although this analyzer is not choosy, mismatched with gradient elution, stretches little sensitivity and is fairly vulnerable to flow rate and temperature changes but still, it has been broadly applied to different matrices, such as cider (Picinelli et al., 2000) or powder milk (Chavez-Servin et al., 2006), meat products (Vendrell-Pascuas et al., 2000), among many others.

To improve sensitivity evaporative-light scattering detectors (ELSD) are used. This sensor is for non-volatile constitutes and gives almost the same behavior for every constituent. Furthermore, since the solvent vanishes just beforehand detection, gradient elution is permissible by ELSD. This detector for the direct analysis of carbohydrates in food has been used with HPLC and a silica gel column (Wei et al., 1997; Wei and Ding, 2002). ELSD has also been used to confirm the occurrence of soluble sugars in onion detection paired to amino columns and size exclusion (Davis et al., 2007). Furthermore, ELSD has proven to be useful as a sensor in mixture with HPAEC separations of carbohydrates (Pierce et al., 2006; Muir et al., 2009).

Although food carbohydrate analysis has many applications of these two sensors, electrochemical detection is the most commonly used method. Specifically, the perfect detector for HPLC partitions of carbohydrates is pulsed amperometry detection (PAD). The compounds that have oxidizable functional groups at the detection voltage active, PAD is appropriate. Essentially, PAD applies a gold electrode and works better in an alkaline medium, which makes it flawlessly appropriate to be paired to HPAEC carbohydrates separations. PAD gives with better sensitivities in comparison to abovementioned RI and ELSD detectors in the examination of underivatized carbohydrates that can spread LODs below 10 μg/L (Cataldi et al., 2000). The potential waveform has sufficiently enhanced carbohydrates discovery by the use of PAD. There are three steps of most widely used waveform, each one lasting a given time (60–300 ms): detection, that includes the direct oxidation of the electrode surface; using higher potential to fully oxidize the electrode surface, cleaning, and regeneration, using a very low potential to expel the former gold-oxide film and consent the electrode ready for the latter detection step. Together, to get the best possible outcomes in each application, potentials, and times intervals have to be set.

The utilization of HPLC with MS detectors also has a significant increment despite the good outcomes gained by using PAD detection in food carbohydrates examination. The usage of HPLC-MS in carbohydrate analysis has many benefits because it can reduce the essential time for separation and purification for the identification of unknown carbohydrates and verify saccharides identification. For the examination of food carbohydrates, MS detection has joined with HPAEC. This pairing is not easy, it provides the technological mismatch that postures the typical alkaline conditions of HPAEC with ionization and volatilization preceding to MS typically utilizing an electrospray interface (ESI). Due to salts' less instability, it renders the creation of electrospray. So desalting device has to be active between the HPAEC column and the ESI-MS detector to work at these conditions. Moreover, in determination to create Li^+ (or chloride) adducts which rise carbohydrates signal in the MS analyzer, a solution of lithium chloride is added to the eluent. Letting the detection of high molecular mass carbohydrates (>2000 Da) with a single quadrupole analyzer (Bruggink et al., 2005) is allowed by using this method in which several carbohydrates of different nature were properly noticed in diverse food and beverages.

For the examination of sugars, HPLC isotope ratio MS (IRMS) can also be joined with HPLC. The presence of ^{13}C in different compounds is to be measured by this technique. Honey and high fructose corn syrup had a widely different natural occurrence of ^{13}C. Therefore, to find the contamination of honey with less quantity of corn syrup (Abramson et al., 2001) this method was useful. By studying the ^{13}C ratios of different sugars, to disclose contaminations in fruit juices, the same method was also used (Guillou et al., 1999).

To find carbohydrates, other detection approaches have also been used to a low amount of joined with HPLC. For example, in food samples, Fourier transforms IR spectrometry (FTIR) has joined with HPLC to find different carbohydrates. By using this process glucose, sucrose, fructose, and maltose were quantified and determined in beverages (Kuligowski et al., 2008).

To be utilized as a C detector inductively attached plasma-atomic emission spectrometry (ICP-AES) has been also joined to HPLC. To prevent interventions in the detection in the moveable phases, the usage of organic compounds has to be evaded. Sucrose, fructose, glucose, sorbitol, and lactose, combine with alcohols and other carboxylic acids, were noticed and counted under these conditions, results attained with UV-V and RI sensors were similar, in terms of LODs (Paredes et al., 2006). In summary, to examine carbohydrates in foods joining different processes and sensors, HPLC has been largely applied (Martinez-Montero et al., 2004; Soga, 2002; PerisTortajada, 2000). These uses involve control of products, control of manufacturing processes, authenticity determination, detection of adulterations, etc (Cataldi et al., 1999; Megherbi et al., 2009).

REFERENCES

1. Aboubacar, A., and Hamaker, B. R., (2000). Low molecular weight soluble starch and its relationship with sorghum couscous stickiness. *J. Cereal Sci., 31*, 119–126.
2. Abramson, F. P., Black, G. E., and Lecchi, P., (2001). Application of high-performance liquid chromatography with isotope-ratio mass spectrometry for measuring low levels of enrichment of underivatized materials. *J. Chromatogr. A, 913*, 269–273.
3. Aliani, M., and Farmer, L. J., (2002). Post column derivatization method for determination of reducing and phosphorylated sugars in chicken by high-performance liquid chromatography. *J. Agric. Food Chem., 50*, 2760–2766.
4. Arias, V. C., Castells, R. C., Malacalza, N., Lupano, C. E., and Castells, C. B., (2003). Determination of oligosaccharide patterns in honey by solid-phase extraction and high-performance liquid chromatography. *Chromatographia, 58*, 797–801.
5. Arya, M., and Mohan, R. L. J., (2007). An impression of coffee carbohydrates. *Crit. Rev. Food Sci. Nutr., 67*, 47–51.
6. Bartolozzi, F., Bertazza, G., Bassi, D., and Cristoferi, G., (1997). Simultaneous determination of soluble sugars and organic acids as their trimethylsilyl derivatives in apricot fruits by gas-liquid chromatography. *J. Chromatogr. A, 758*, 99–107.
7. Bruggink, C., Maurer, R., Herrmann, H., Cavalli, S., and Hoefler, F., (2005). Analysis of carbohydrates by anion exchange chromatography and mass spectrometry. *J. Chromatogr. A., 1085*, 104–109.
8. Careri, M., Bianchi, F., and Corradini, C., (2002). Recent advances in the application of mass spectrometry in the food-related analysis. *J. Chromatogr. A, 970*, 3–64.
9. Castellari, M., Versari, A., Spinabelli, U., Galassi, S., and Amati, A., (2000). An improved HPLC method for the analysis of organic acids, carbohydrates, and alcohol in grape musts and wines. *J. Liq. Chromatogr. Related Technol., 23*, 2047–2056.
10. Cataldi, T. R. I., Angelotti, M., and Bufo, S. A., (1999). Method development for the quantitative determination of lactulose in heat-treated milks by HPAEC with pulsed amperometric detection. *Anal. Chem., 71*, 4919–4925.
11. Cataldi, T. R. I., Campa, C., and De Benedetto, G. E., (2000).

Carbohydrate analysis by high-performance anion-exchange chromatography with pulsed amperometric detection, the potential is still growing. *Fresenius J. Anal. Chem., 368*, 739–758.

12. Chandra, K., Ghosh, K., Ojha, A. K., and Islam, S. S., (2009). A protein containing glucan from an edible mushroom. Termitomyces microcarpus. *Nat. Prod. Commun., 4*, 553–556.
13. Chavez-Servin, J. L., Romeu-Nadal, M., Castellote, A. I., and Lopez-Sabater, M. C., (2006). Evolution of free mono- and di-saccharide content of milk-based formula powder during storage. *Food Chem., 97*, 103–108.
14. Cifuentes, A., (2009). Food analysis and foodomics. *J. Chromatogr. A, 1216*, 7109.
15. Cmelik, R., Stikarovska, M., and Chmelik, J., (2004). Different behavior of dextrans in positive-ion and negative-ion mass spectrometry. *J. Mass Spectrom., 39*, 1467–1473.
16. Corsaro, A., Chiacchio, U., Pistara, P., and Romeo, G., (2004). Microwave-assisted chemistry of carbohydrates. *Current Org. Chem., 8*, 511–538.
17. Cortacero-Ramirez, S., Segura-Carretero, A., Cruces-Blanco, C., Hernainz-Bermudez, M., and Fernandez-Gutierrez, A., (2004). Analysis of carbohydrates in beverages by capillary electrophoresis with precolumn derivatization and UV detection. *Food Chem., 87*, 471–476.
18. Corzo-Martinez, M., Moreno, F. J., Villamiel, M., and Harte, F. M., (2010). Characterization and improvement of rheological properties of sodium caseinate glycated with galactose, lactose, and dextran. *Food Hydrocolloids, 24*, 88–97.
19. Cotte, J. F., Casabianca, H., Chardon, S., Lheritier, J., and Grenier-Loustalot, M. F., (2004). Chromatographic analysis of sugars applied to the characterization of monofloral honey. *Anal. Bioanal. Chem., 380*, 698–705.
20. Cotte, J. F., Casabianca, H., Chardon, S., Lheritier, J., and Grenier-Loustalot, M. F., (2003). Application of carbohydrate analysis to verify honey authenticity. *J. Chromatogr. A, 1021*, 145–155.
21. Cui, S. W., (2005). *Food Carbohydrates, Chemistry, Physical Properties and Applications*. Boca Raton, CRC Press., *1*, 10-44.
22. Davis, F., Terry, L. A., Chope, G. A., and Faul, C. F. J., (2007). Effect

of extraction procedure on measured sugar concentrations in Onion (*Allium cepa* L.) bulbs. *J. Agric Food Chem., 55*, 4299–4306.

23. de la Fuente, E., Sanz, M. L., Martinez-Castro, I., Sanz, J., and Ruiz-Matute, A. I., (2007). Volatile and carbohydrate composition of rare unifloral honeys from Spain. *Food Chem., 105*, 84–93.
24. Dourado, F., Vasco, P., Barros, A., Mota, M., Coimbra, M. A., and Gama, F. M., (2003). Characterization of Chilean hazelnut (*Gevuina avellana*) tissues, light microscopy and cell wall polysaccharides. *J. Sci. Food Agric, 83*, 158–165.
25. Eremeeva, T., (2003). Size-Exlusion chromatography of enzimatically treated cellulose and related polysaccharides, a review. *J. Biochem. Biophys. Methods, 56*, 253–264.
26. Fuzfai, Z., and Molnar-Perl, I., (2007). Gas chromatographic-mass spectrometric fragmentation study of flavonoids as their trimethylsilyl derivatives, analysis of flavonoids, sugars, carboxylic, and amino acids in model systems and in citrus juices. *J. Chromatogr. A, 1149*, 88101.
27. Fuzfai, Z., Katona, Z. F., Kovacs, E., and Molnar-Perl, I., (2004). Simultaneous identification and quantification of the sugar, sugar alcohol and carboxylic acid contents of sour cherry, apple, and ber fruits as their trimethylsilyl derivatives by gas chromatography-mass spectrometry. *J. Agric. Food Chem., 52*, 7444–7452.
28. García-Baños, J. L., Olano, A., and Corzo, N., (2002). Changes in the carbohydrate fraction during manufacture and storage of enteral formulas. *J Food Sci., 67*, 3232–3235.
29. Garcia-Cañas, V., and Cifuentes, A., (2008). Recent advances in the application of capillary electromigration methods for food analysis. *Electrophoresis, 29*, 294–309.
30. Geisser, A., Hendrich, T., Boehm, G., and Stahl, B., (2005). Separation of lactose from human milk oligosaccharides with simulated moving bed chromatography. *J. Chromatogr. A, 1092*, 17–23.
31. Gibson, P. R., Newham, E., Barrett, J. S., Shepherd, S. J., and Muir, J. G., (2007). Fructose malabsorption and the bigger picture. *Aliment. Pharmaciol. Ther., 25*, 349–363.
32. Gniechwitz, D., Reichardt, N., Ralph, J., Blaut, M., Steinhart, H., and Bunzel, M., (2008). Isolation and characterisation of a coffee melanodin fraction. *J. Sci. Food Agric., 88*, 2153–2160.
33. Godshall, M. A., Roberts, E. J., and Miranda, X. M., (2001).

Composition of the soluble, nondyalizable components in raw cane sugar. *J. Food Process. Preservation, 25*, 323–335.

34. Guillou, C., Koziet, J., Rossmann, A., and Martin, G. J., (1999). Determination of the 13C contents of organic acids and sugars in fruit juices, an inter-comparison study. *Anal. Chim. Acta, 388*, 137–143.
35. Hartmann, G., Piber, M., and Koehler, P., (2005). Isolation and chemical characterization of water-extractable arabinoxylans from wheat and rye during breadmaking. *Eur. Food Res. Technol., 221*, 487–492.
36. Harvey, D. J., (2008). Analysis of carbohydrates and glycoconjugates by matrix-assisted laser desorption/ionization mass spectrometry, An update covering the period (2001)–(2002). *Mass Spectrom. Rev., 27*, 125–201.
37. Haska, L., Nyman, M., and Andersson, R., (2008). Distribution and characterization of fructan in wheat milling fractions. *J. Cereal Sci., 48*, 768–774.
38. Haslam, S. M., North, S. J., and Dell, A., (2006). Mass spectrometric analysis of N and O-glycosylation of tissues and cells. *Current Opin. Struct. Biol., 16*, 584–591.
39. Hernandez, O., Ruiz-Matute, A. I., Olano, A., Moreno, F. J., and Sanz, M. L., (2009). Comparison of fractionation techniques to obtain prebiotic galactooligosaccharides. *Int. Dairy J., 19*, 531536.
40. Herrero, M., García-Cañas, V., Simo, C., and Cifuentes, A., (2009). Recent advances in the application of capillary electromigration methods for food analysis and foodomics. *Electrophoresis, 1*, 1-30.
41. Hounsome, N., Hounsome, B., Tomos, D., and Edwards-Jones, G., (2008). Plant metabolites and nutritional quality of vegetables. *J. Food Sci., 73*, R48–R65.
42. Hsu, K. L.,and Mahal, L. K., (2009). Sweet tasting chips, microarray-based analysis of glycans. *Current Opin. Chem. Biol., 13*, 427–432.
43. Ibañez, E., and Cifuentes, A., (2003). New analytical techniques in food science. *Crit. Rev. Food Sci., 41*, 413–450.
44. Jager, A. V., Tonin, F. G., and Tavares, M. F. M., (2007). Comparative evaluation of extraction procedures and method validation for determination of carbohydrates in cereals and dairy products by capillary electrophoresis. *J. Sep. Sci., 30*, 586–594.
45. Kakita, H., Kamishima, H., Komiya, K., and Kato, Y., (2002). Simultaneous analysis of monosaccharides and oligosaccharides

by high-performance liquid chromatography with postcolumn fluorescence derivatization. *J. Chromatogr. A, 961*, 77–82.

46. Kamada, T., Nakajima, M., Nabetani, H., and Iwamoto, S., (2002). Pilot-scale study of the purification and concentration of non-digestible saccharides from Yacon Rootstock using membrane technology. *Food Sci. Technol. Res.*, *8*, 172–177.
47. Khandurina, J., and Guttman, A., (2005). High resolution capillary electrophoresis of oligosaccharide structural isomers. *Chromatographia, 62*, S37–S41.
48. Kim, D., Park, C. S., and Park, K. H., (2003). Preparation and characterization of maltosyl-sucrose isomers produced by transglycosylation of maltogenic amylase from *Bacillus stearothermophilus*. *J. Mol. Catal. B-Enzymatic.*, *26*, 293–305.
49. Kim, H. S., Yi, S. Y., Kim, J., et al. (2009b). Novel application of surface plasmon resonance biosensor chips for measurement of advanced glycation end products in serum of Zucker diabetic fatty rats. *Biosens. Bioelectron.*, *25*, 248–252.
50. Kim, Y. P., Park, S., Oh, E., Oh, Y. H., and Kim, H. S., (2009b). On-chip detection of protein glycosylation based on energy transfer between nanoparticles. *Biosens. Bioelectron., 24*, 1189–1194.
51. Kitahara, K., and Copeland, L., (2004). A simple method for fractionating debranched starch using a solid reversed-phase cartridge. *J. Cereal Sci., 39*, 91–98.
52. Klampfl, C. W., and Buchberger, W., (2001). Determination of carbohydrates by capillary electrophoresis with electrospray-mass spectrometric detection. *Electrophoresis, 22*, 2737–2742.
53. Kolhed, M., and Larlberg, B., (2005). Capillary electrophoretic separation of sugars in fruit juices using on-line mid-infrared Fourier transform detection. *Analyst, 130*, 772–778.
54. Kuligowski, J., Quintas, C., Garrigues, S., and de la Guardia, M., (2008). On-line gradient liquid chromatography-Fourier transform infrared spectrometry determination of sugars in beverages using chemometric background correction. *Talanta, 77*, 779–785.
55. La Torre, G., La Pera, L., Rando, R., et al. (2008). Classification of Marsala wines according to their polyphenol, carbohydrate, and heavy metal levels using canonical discriminant analysis. *Food Chem., 110*, 729–734.

56. Lamari, F. N., Kuhn, R., and Karamanos, N. K., (2003). Derivatization of carbohydrates for chromatographic, electrophoretic, and mass spectrometric structure analysis. *J. Chromatogr. B 793*, 15–36.
57. Lastovickova, M., Chmelik, J., and Bobalova, J., (2009). The combination of simple MALDI matrices for the improvement of intact glycoproteins and glycans analysis. *Int. J. Mass Spectrom., 281*, 82–88.
58. Lehtonen, P., and Hurme, R., (1994). Liquid-chromatographic determination of sugars in beer by evaporative light-scattering detection. *J. Inst. Brew., 100*, 343–346.
59. Lion, N., Rossier, J. S., and Girault, H. H., (2006). *Micro Fluidic Applications in Biology, from Technologies to Systems Biology,* Weinheim, Wiley-VCH, *1*, 1-43.
60. Louw, S., Pereira, A. S., Lynen, F., Hanna-Brown, M., and Sandra, P., (2008). Serial coupling of reversed-phase and hydrophilic interaction liquid chromatography to broaden the elution window for the analysis of pharmaceutical compounds. *J. Chromatogr. A, 1208*, 90–94.
61. Mabel, M. J., Sangeetha, P. T., Platel, K., Srinivasan, K., and Prapulla, S. G., (2008). Physicochemical characterization of fructooligosaccharides and evaluation of their suitability as a potential sweetener for diabetics. *Carbohydrate Res., 343*, 55–66.
62. Martinez, M. C., Rodriguez, D. M. C., Guillén, S. D. A., and Barroso, C. G., (2004). Analysis of low molecular weith carobohydrates in food and beverages, a review. *Chromatographia, 59*, 15–30.
63. Martinez-Villaluenga, C., Cardelle-Cobas, A., Olano, A., Corzo, N., Villamiel, M., and Jimeno, M. L., (2008). Enzymatic synthesis and identification of two trisaccharides produced from lactulose by transgalactosylation. *J. Agric. Food Chem., 56*, 557–563.
64. Megherbi, M., Herbretau, B., Dessalces, G., and Grenier-Loustalot, M. F., (2008). Solid phase extreaction of oligo- and polysaccharides; application to malodextrins and honey qualitative analysis. *J. Liq. Chromatogr. Rel. Technol., 31*, 1033–1046.
65. Megherbi, M., Herbreteau, B., Faure, R., and Salvador, A., (2009). Polysaccharides as a marker for detection of corn sugar syrup addition in honey. *J. Agric. Food Chem., 57*, 2105–2111.
66. Mendiola, J. A., Herrero, M., Cifuentes, A., and Ibáñez, E., (2007). Use of compressed fluids for sample preparation, food applications. *J. Chromatogr. A, 1152*, 234–246.

67. Molnar-Perl, I., (2000). Role of chromatography in the analysis of sugars, carboxylic acids and amino acids in food. *J. Chromatogr. A, 891*, 1–32.
68. Momenbeik, F., Johns, C., Breadmore, M. C., Hilder, E. F., Macka, M., and Haddad, P. R., (2006). Sensitive determination of carbohydrates labelled with p-nitroaniline by capillary electrophoresis with photometric detection using a 406 nm light-emitting diode. *Electrophoresis, 27*, 4039–4046.
69. Montañes, F., Fornari, T., Martin-Alvarez, P. J., Corzo, N., Olano, A., and Ibañez, E., (2006). Selective recovery of tagatose from mixtures with galactose by direct extraction with supercritical CO_2 and different cosolvents. *J. Agric. Food Chem., 54*, 8340–8345.
70. Montañes, F., Fornari, T., Martin-Alvarez, P. J., et al. (2007). Selective fractionation of disaccharide mixtures by supercritical CO_2 with ethanol as co-solvent. *J. Supercrit. Fluids, 41*, 61–67.
71. Montilla, A., Moreno, F. J., and Olano, A., (2005). A reliable gas capillary chromatographic determination of lactulose in dairy samples. *Chromatographia, 62*, 311–314.
72. Montilla, A., Ruiz-Matute, A. I., Sanz, M. L., Martinez-Castro, I., and Del Castillo, M. D., (2006b). Difructose anhydrides as quality markers of honey and coffee. *Food Res. Int., 39*, 801–806.
73. Montilla, A., Van de Langemaat, J., Olano, A., and Del Castillo, M. D., (2006a). Determination of oligosaccharides by conventional high resolution gas chromatography. *Chromatographia, 63*, 453–458.
74. Morales, V., Corzo, N., and Sanz, M. L., (2008). HPAEC-PAD oligosaccharide analysis to detect adulterations of honey with sugar syrups. *Food Chem., 107*, 922–928.
75. Morales, V., Olano, A., and Corzo, N., (2004). Ratio of maltose to maltulose and furosine as quality parameters for infant formula. *J. Agric. Food Chem., 52*, 6732–6736.
76. Morales, V., Sanz, M. L., Olano, A., and Corzo, N., (2006). Rapid separation on activated charcoal of high oligosaccharides in honey. *Chromatographia, 64*, 233–238.
77. Morales-Cid, G., Simonet, B. M., Cardenas, S., and Valvarcel, M., (2007). On-capillary sample cleanup method for the electrophoretic determination of carbohydrates in juice samples. *Electrophoresis, 28*, 1557–1563.

78. Muir, J. G., Rose, R., Rosella, O., et al. (2009). Measurement of short-chain carbohydrates in common Australian vegetables and fruits by high-performance liquid chromatography (HPLC). *J. Agric. Food Chem. 57*, 554–565.

79. Muir, J. G., Shepherd, S. J., Rosella, O., Rose, R., Barrett, J. S., and Gibson, P. R., (2007). Fructan and free fructose content of common Australian vegetables and fruit. *J. Agric. Food Chem., 55*, 6619–6627.

80. Mullin, W. J., and Emmons, D. B., (1997). Determination of organic acids and sugars in cheese, milk, and whey by high-performance liquid chromatography. *Food Res. Int., 30*, 147–151.

81. Murphy, S. P., and Johnson, R., (2003). The scientific basis of recent US guidance on sugars intake. *Am. J. Clin. Nutr., 78*, 827S–833S.

82. Ninonuevo, M. R., Park, Y., Yin, H., et al. (2006). A strategy for annotating the human milk glycome. *J. Agric. Food Chem., 54*, 7471–7477.

83. Ninonuevo, M., An, H., Yin, H., et al. (2005). Nanoliquid chromatography-mass spectrometry of oligosaccharides employing graphitized carbon chromatography on microchip with a high accuracy mass analyzer. *Electrophoresis, 26*, 3641–3646.

84. Noe, C. R., Lachmann, B., Mollenbeck, S., and Richter, P., (1999). Determination of reducing sugars in selected beverages by capillary electrophoresis. *Z. Lebensm. Unters. Forsch A, 208*, 148–152.

85. Okatch, H., Torto, N., and Armateifio, J., (2003). Characterization of legumes by enzymatic hydrolysis, microdialysis sampling, and micro-high-performance anion-exchange chromatography with electrospray ionization mass spectrometry. *J. Chromatogr. A, 992*, 67–74.

86. Paredes, E., Maestre, S. E., Prats, S., and Todoli, J. L., (2006). Simultaneous determination of carbohydrates, carboxylic acids, alcohols, and metals in foods by high-performance liquid chromatography inductively coupled plasma atomic emission spectrometry. *Anal. Chem., 78*, 6774–6782.

87. Park, Y. K., and Yetley, E. A., (1993). Intakes and food sources of fructose in the United States. *Am. J. Clin. Nutr., 58*, 737S–747S.

88. Peris-Tortajada, M., (2000). HPLC determination of carbohydrates in foods. In: Nollet, L. M. L., (ed.) *Food Analysis by HPLC,* New York, Marcel Dekker, 1-29.

89. Picinelli, A., Suarez, B., Moreno, J., Rodriguez, R., Caso-García, L.

M., and Mangas, J. J., (2000). Chemical characterization of asturian cider. *J. Agric. Food Chem., 48*, 3997–4002.

90. Pierce, K. M., Hope, J. M., Hoggard, J. C., and Synovec, R. E., (2006). A principal component analysis based method to discover chemical differences in comprehensive two-dimensional gas chromatography with time-of-flight mass spectrometry (GC×GC-TOFMS) separations of metabolites in plant samples. *Talanta, 70*, 797–804.
91. Rojas-Escudero, E., Alarcon-Jimenez, A. L., Elizalde-Galvan, P., and Rojo-Callejas, F., (2004). Optimization of carbohydrate silylation for gas chromatography. *J. Chromatogr. A.*, *1027*, 117–120.
92. Rovio, S., Yli-Kauhaluoma, J., and Siren, H., (2007). Determination of neutral carbohydrates by CZE with direct UV detection. *Electrophoresis, 28*, 3129–3135.
93. Ruiz-Matute, A. I., Montilla, A., del Castillo, M. D., Martinez-Castro, I., and Sanz, M. L., (2007c). A GC method for simultaneous analysis of bornesitol, other polyalcohols and sugars in coffee and its substitutes. *J. Sep. Sci., 30*, 557–562.
94. Ruiz-Matute, A. I., Ramos, L., Martinez-Castro, I., Sanz, M. L., (2008). Fractionation of honey carbohydrates using pressurized liquid extraction with activated charcoal. *J. Agric. Food Chem., 56*, 8309–8313.
95. Ruiz-Matute, A. I., Sanz, M. L., and Martinez-Castro, I., (2007b). Use of gas chromatography-mass spectrometry for identification of a new disaccharide in honey. *J. Chromatogr. A, 1157*, 480–483.
96. Ruiz-Matute, A. I., Sanz, M. L., Corzo, N., et al. (2007a). Purification of lactulose from mixtures with lactose using pressurized liquid extraction with ethanol-water at different temperatures. *J. Agric. Food Chem., 55*, 3346–3350.
97. Sharon, N., (2008). Lectins, past, present, and future. *Biochem. Soc. Trans., 36*, 1457–1460.
98. Soga, T., (2002). Analysis of carbohydrates in food and beverages by HPLC and CE. In: El Rassi, Z., (ed.), *Carbohydrate Analysis by Modern Chromatography and Electrophoresis*, 483503. Amsterdam, Elsevier Science.
99. Soria, A. C., Sanz, M. L., and Villamiel, M., (2009). Determination of minor carbohydrates in carrot (*Daucus carota L.*) by GC-MS. *Food Chem., 114*, 758–762.

100. Stikarovska, M., and Chmelik, J., (2004). Determination of neutral oligosaccharides in vegetables by matrix-assisted laser desorption/ionization mass spectrometry. *Anal. Chim. Acta, 520*, 47–55.

101. Suzuki, S., Kelly, J. F., Locke, S. J., Thibault, P., and Honda, S., (2003). Derivatization of Carbohydrates. In: Thibault, P., and Honda, S., (eds.), *Capillary Electrophoresis of Carbohydrate,* New York, Humana Press., *1,* 1-44

102. Taga, A., Suzuki, S., and Honda, S., (2001). Capillary electrophoretic analysis of carbohydrates derivatized by in-capillary condensation with 1-phenyl-3-methyl-5-pyrazolone. *J. Chromatogr. A, 911*, 259–267.

103. Takahashi, N., and Kato, K., (2003). GALAXY (Glycoanalysis by the three axes of MS and chromatography), a web application that assists structural analyses of N-glycans. *Trends Glycosci. Glycotech., 15*, 235–251.

104. Takahashi, N., Nakagawa, H., Fujikawa, K., Kawamura, Y., and Tomiya, N., (1995). Threedimensional elution mapping of pyridylaminated N-linked neutral and sialyl oligosaccharides. *Anal. Biochem., 226*, 139–146.

105. Terrab, A., Vega-Perez, J. M., Diez, M. J., and Heredia, F. J., (2001). Characterization of northwest Moroccan honeys by gas chromatographic-mass spectrometric analysis of their sugar components. *J. Sci. Food Agric., 82*, 179–185.

106. Tomiya, N., Awaya, J., Kurono, M., Endo, S., Arata, Y., and Takahashi, N., (1988). Analyses of N-linked oligosaccharides using a two-dimensional mapping technique. *Anal. Biochem., 171*, 73–90.

107. Torto, N., Hofte, A., Van, D. H. R., et al. (1998). Microdialysis introduction high-performance anion-exchange chromatography/ion spray mass spectrometry for monitoring of on-line desalted carbohydrate hydrolysates. *J. Mass Spectrom., 33*, 334–341.

108. Torto, N., Lobelo, B., and Gorton, L., (2000). Determination of saccharides in wastewater from the beverage industry by microdialysis sampling, microbore high performance anion exchange chromatography and integrated pulsed electrochemical detection. *Analyst., 125*, 1379–1381.

109. Vendrell-Pascuas, S., Castellote-Bargallo, A. I., and Lopez-Sabater, M. C., (2000). Determination of inulin in meat products by high performance liquid chromatography with refractive index detection. *J.*

Chromatogr. A, 881, 591–597.

110. Ward, R. M., Gao, Q., de Bruyn, H., Gilbert, R. G., and Fitzgerald, M. A., (2006). Improved methods for the structural analysis of the amylase-rich fraction from rice flour. *Biomacromolecules, 7*, 866–876.
111. Wei, Y., and Ding, M. Y., (2002). Ethanolamine as modifier for analysis of carbohydrates in foods by HPLC and evaporative light scattering detection. *J. Liq. Chromatogr. Rel. Technol., 25*, 1769–1778.
112. Wei, Y., Hendrix, D. L., and Nieman, R., (1997). Diglucomelezitose, a novel pentasaccharide in silver leaf whitefly honeydew. *J. Agric. Food Chem., 45*, 3481–3486.
113. Weston, R. J., and Brocklebank, L. K., (1999). The oligosaccharide composition of some New Zealand honeys. *Food Chem., 64*, 33–37.
114. Xiaoli, X., Liyi, Y., Shuang, H., et al. (2008). Determination of oligosaccharide contents in 19 cultivars of chickpea (*Cicer arietinum L*) seeds by high performance liquid chromatography. *Food Chem., 111*, 215–219.
115. Yagi, H., and Kato, K., (2009). Multidimensional HPLC mapping method for the structural analysis of anionic N-glycans. *Trends Glycosci. Glycotech., 21*, 95–104.
116. Yagi, H., Takahashi, N., Yamaguchi, Y., et al. (2005). Development of structural analysis of sulfated N-glycans by multidimensional high performance liquid chromatography mapping methods. *Glycobiology, 15*, 1051–1060.
117. Zaia, J., (2004). Mass spectrometry of oligosaccharides. *Mass Spectrom. Rev., 23*, 161–227.
118. Zamfir, A. D., Bindila, L., Lion, N., Allen, M., Girault, H. H., and Peter-Katalinic, J., (2005). Chip electrospray mass spectrometry for carbohydrate analysis. *Electrophoresis, 26*, 3650–3657.
119. Zhai, C., Li, C., Qiang, W., Xei, J., Yu, X., and Ju, H., (2007). Amperometric detection of carbohydrates with a portable silicone/quartz capillary microchip by designed fracture sampling. *Anal. Chem., 79*, 9427–9432.
120. Zhang, L., Xu, J., Zhang, L., Zhang, W., and Zhang, Y., (2003). Determination of 1-phenyl-3-methyl-5-pyrazolone-labeled carbohydrates by liquid chromatography and micellar electrokinetic chromatography. *J. Chromatogr. B, 793*, 159–165.

Chapter 7

Chemical Analysis of Lipids

CONTENTS

7.1. INTRODUCTION

Among the components of foods, lipids are the major ones and for a lot of reasons, they are important for our diet. They provide necessary lipid nutrients and are the main source of energy (Billing et al., 1953; Hopmans et al., 2000). However, excessive-consumption of some particular lipid nutrients can be harmful to one's health, for example, saturated fats and cholesterol. Lipid components have a key role in a lot of foods and determine on the whole physical characteristics, such as appearance, texture, and flavor (Karabelnik and Zbinden, 1975; Gbogouri et al., 2006). It is hard to develop substitutes for many foods which are low-fat just for this reason because some of the mainly essential physical characteristics are disappearing once the fat is removed. To conclude, a lot of fats are liable to oxidation of lipid, which results in the creation of the potentially injurious products and off-flavors (Evershed et al., 1995; Byrdwell, 2001; Cai and Syage, 2006). A few of the many important characteristics of concern to food analyst are:

1. Type of lipids existing in a particular food;
2. The total concentration of lipids;
3. Structural organization of lipids in food; and
4. Physicochemical characteristics of lipids, for instance, melting point, crystallization, rheology, smoke point, color, and density.

7.2. PROPERTIES OF LIPIDS IN FOODS

Lipids are generally defined as those elements that are insoluble in water, but soluble in organic solvents (as chloroform, hexane or ether) (Passi et al., 1981; Henson et al., 1985; Stübiger and Belgacem, 2007). These substances include monoacylglycerol, diacylglycerols, and triacylglycerols, acids, phospholipids, vitamins A and D, free fatty, carotenoids, and sterols (Fennema et al., 1994; Yang and Paulson, 2000). Hence, the lipid portion of a fatty food comprises of a complicated mixture of various types of molecules. Even so, the main components of many foods are triacylglycerols, generally making more than 95 to 99% of the whole lipids present. Triacylglycerols are esters of a glycerol molecule and three fatty acids (Kinsella, 1981; Nichols and Sanderson, 2003). The fatty acids usually present in food differ in position on the molecule of glycerol, degree of unsaturation and length of chain. Hence, triacylglycerol fraction itself is composed of a complicated combination of different types of molecules. The accurate nature of lipids physiochemical and nutritional properties are determined by the different profile of lipids of each kind of fat present in it. Food scientists repeatedly use the terms lipid, oil, and fat interchangeably. Even though sometimes lipids are liquid at a particular temperature are called oil, whereas lipids that are solid at the particular temperature are called fat (Ali et al., 2013; Michalski et al., 2013).

7.3. SAMPLE SELECTION AND PRESERVATION

Just like in the case of any analytical method, the credibility of the results is dependent on accurate sampling as well as the preservation of the sample before its analysis. Preferably, the sample whose composition is being examined should describe as accurately as possible the food from which it is taken (Windham et al., 1989; Pfeiffer and Varney, 2002). The model preparation we need in the analysis of lipid is based on the nature of the component of lipid (e.g., physical state, susceptibility to oxidation, volatility), the form of analytical procedure used (e.g., non-solvent instrumental or extraction, solvent extraction) and the type of food analysis (e.g., milk, cookie, margarine, meat, dairy cream). To make the decision of the most suitable procedure for sample preparation, it is essential to know the position of the principal lipids in the food and its physical structure. It is essential to utilize different procedures for different kinds of foods considering the unique nature of each food (Nickerson and Lynch, 2001; Noonan, 2007; Wolf et al., 2011). Official methods, that specify to follow the accurate sample preparation procedure, have been developed for particular types of foods. Simply, the preparation of the sample should be made using an environment that reduces any alteration in the properties of the lipid fraction. To avoid lipid oxidation, it is essential to preserve the lipid sample by employing cold temperatures, nitrogen environment, low light intensity, and additional antioxidants. It may be compulsory to cautiously handle the sample and manage the temperature if the crystal structure or solid fat content (SFC) is important (Dalzell and Shelton, 1997; Liu et al., 2013; Tarricone et al., 2015).

7.4. DETERMINATION OF TOTAL LIPID CONCENTRATION

It is essential to precisely determine the overall content of fat in foods due to many reasons such as (Frings and Dunn, 1970; Smedes, 1999):

1. Legal (to conform to nutritional labeling laws and identity standards);
2. Economic (expensive ingredients are not given away);
3. Processing (processing conditions mainly rely on the total lipid content);
4. Health (development of low-fat foods); and
5. Quality (properties of food depend on the overall lipid content).

The key physicochemical properties used to differentiate the lipids (analyte) from other constituents of the foods (matrix) include spectroscopic properties, physical characteristics (e.g., comparatively low density), immiscibility with water and solubility in organic solvents (Strasberg et al., 1990; Donovan et al., 1991). The analytical techniques depending on these principles can be easily categorized into these different types: (i) instrumental methods; (ii) non-solvent extraction and (iii) solvent extraction (Frings et al., 1972; Nourooz-Zadeh et al., 1996).

7.4.1. Solvent Extraction

Food analysts can use the convenient technique of separating the water-soluble components such as minerals, carbohydrates, and proteins from lipid components present in foods. This can be achieved by knowing the fact that lipids are insoluble in water but soluble in organic solvents (Dodds et al., 2004; Young et al., 2010; Teo and Idris, 2014). Solvent extraction techniques are amongst the most generally used techniques for fat content determination and lipid isolation from foods (Schäfer, 1998; Long and Abdelkader, 2011; Grima et al., 2013).

7.4.1.1. Sample Preparation

Many steps are usually involved in sample preparation for the extraction of solvent (Galliard, 1968; McLemore et al., 1998; Koivusalo et al., 2001):

1. *Drying sample.* It is usually compulsory to dry the samples before the extraction of the solvent. This step is necessary to assist the penetration of organic solvents by removing the water molecules, otherwise, inefficient extraction would be observed.

2. *Particle size reduction.* To get more surface area of lipid exposed to the solvent and to produce a further homogeneous sample, dried samples are often finely ground before the solvent extraction. To decrease the tendency for the oxidation of lipid to take place, grinding is mostly implemented at low temperatures.

3. *Acid hydrolysis.* Some foods have lipids that are complexed with glycolipids (polysaccharides) or lipoproteins (proteins). It is necessary that the bonds that hold the non-lipid and lipid components should be broken together prior to solvent extraction. This will assist in the concentration determination of the constituents. To release the bound lipids into a form in which they are easily extractable, acid hydrolysis is used commonly, e.g., if a sample is heated for 1 hour in the existence of 3NHCl acid, it is digested.
4. *Solvent Selection.* The *perfect* solvent for the extraction of a lipid would entirely extract each lipid component present in the food, while all the other components will be left behind. In reality, the solvent extraction efficiency mainly depends on the polarity of present lipids rather than the polarity of the used solvent. Polar lipids (e.g., phospholipids or glycolipids) are more soluble in polar solvents (e.g., alcohols) than in non-polar solvents (e.g., hexane). Alternatively, non-polar lipids (e.g., triacylglycerols) are less soluble in polar solvents than in non-polar ones. It is not possible to select one organic solvent for the extraction of them all by the fact that there are different polarities of different lipids. Therefore the lipid content determined by the extraction of solvent depends on the characteristics of the organic solvent utilized for carrying out the extraction: the overall lipid content determined utilizing one solvent can be dissimilar from the content determined to employ another solvent. Moreover, a solvent should have a relatively low boiling point (easily removable by evaporation), inexpensive, be nonflammable (safety reasons) and be non-toxic. It is quite hard to get a single solvent which fulfills all of these requirements. Petroleum ether and Ethyl ether are the most generally used solvents, but hexane and pentane are also utilized for a few foods.

7.4.1.2. Batch Solvent Extraction

These techniques depend on mixing the solvent and the sample in an appropriate container, such as a separatory funnel. The aqueous phase and organic solvent are separated (either by centrifugation or gravity) by shaking the container vigorously. The aqueous phase is decanted off, and then by the evaporation of the solvent and measurement of the mass of remaining lipid, the lipid concentration in the solvent is determined.

% Lipid = 100 x (M_{Lipid}/M_{Sample})

To enhance the efficiency of the process of extraction one may have to repeat the procedure many times. In the following case, the aqueous phase would go through more extractions by fresh solvent, after that every solvent fractions would be collected and the lipid determined by measuring after solvent evaporation. The efficiency of the extraction of a specific type of lipid by a specific type of solvent can be calculated by an equilibrium partition coefficient,

$K = C_{solven}t/C_{aqueous}$

Where, $C_{solvent}$ is the concentration of lipids in the solvent; $C_{aqueous}$ is the concentration of lipids in the aqueous phase.

The efficiency of the extraction process depends directly on the partition coefficient.

7.4.1.3. Semi-Continuous Solvent Extraction

To enhance the efficiency of extraction of lipids from foods, semi-continuous solvent extraction methods are used commonly. The most frequently used example of a semi-continuous method is the Soxhlet method. In this method, firstly a sample is dried, after that, it is grounded into small particles and then it is placed in a porous thimble. That thimble is placed in an extraction chamber, that chamber is hanging below a condenser and above a flask having the solvent. The solvent is evaporated by heating the flask. Then the solvent is converted into a liquid by going up into the condenser and that trickles in the extraction chamber having the sample. Eventually, the solvent is collected in the extraction chamber and surrounds the sample. Whenever the solvent around the sample goes beyond a specific level it overflows and drops down back into the boiling flask because the extraction chamber is designed so. The solvent extracts the lipids and carries them into the flask as it passes through the sample. Because of the low volatility, the lipids stay in the flask. The final stage of the extraction process is typical of a few hours. At this stage the flask having the lipid and solvent is removed, from the mixture solvent is evaporated and the mass of remaining lipid is measured (M_{Lipid}). For the calculation of the percentage of lipid in the initial sample (M_{Sample}) we can use:

% Lipid = 100 x (M_{lipid}/M_{sample})

Modified versions of the Soxhlet method designed by a lot of instrument manufacturers. These versions can be used to find the total lipid content rapidly and more easily (e.g., Soxtec).

7.4.1.4. Continuous Solvent Extraction

Unlike the Soxhlet method, the extraction chamber of the Goldfish method is designed in such a way that the solvent trickles through the sample instead of building up around it. This decreases the time required to perform the extraction, however, it has a drawback that channeling of the solvent can happen, i.e., the extraction is inefficient as the solvent might preferably take some specific routes through the sample. This is not an issue in the Soxhlet method because the solvent always surrounds the sample.

7.4.1.5. Accelerated Solvent Extraction

To increase the efficiency of solvent extraction it should be carried out at a higher pressure and temperature than normal use. While extracting lipids from food, the effectiveness of a solvent increases by the increase in temperature, but to keep the solvent in the liquid state, the pressure should also be increased. By this method the quantity of solvent required to do the analysis reduces. It is beneficial from an environmental and cost standpoint. At high pressures and temperatures, there are special instruments to do solvent extraction.

7.4.2. Extraction Methods of Non-Solvent Liquid

For the separation of lipids from the rest of the food, many liquid extraction methods use other chemicals instead of organic solvents. For finding the lipid content of milk and a few other dairy products, non-solvent liquid extraction methods like Detergent, Gerber, and Babcock are used.

7.4.2.1. Babcock Method

A particular amount of milk is precisely placed into a special flask (called a Babcock bottle). Sulfuric acid dissolved in milk, digests the protein, releases heat, and breaks down the globule membrane of fat hence, releasing the fat which surrounds the droplets. When the sample is hot (55–60°C) it is centrifuged which in return raises the liquid fat to the neck of the Babcock bottle. The neck is graduated in such a way that it gives the milk fat present in wt%. This method requires almost 45 minutes to complete, and its precision is 0.1%. Phospholipids in milk are in the aqueous phase or at the boundary between the aqueous phases and lipids. That's why this method does not find out phospholipids in milk.

7.4.2.2. Gerber Method

The difference in this method and the Babcock method is that a mixture of isoamyl alcohol and sulfuric acid is used, and the shape of the bottle is slightly different. It is simpler and faster to perform as compared to the Babcock method. Charring of the sugars by heat and sulfuric acid is prevented by the isoamyl alcohol which can be an issue in the Babcock method as it is difficult to read the content of fat from the graduated flask. This method is mostly used in Europe, while on the other hand the Babcock method is mostly used in the USA. It does not determine phospholipids like with the Babcock method.

7.4.2.3. Detergent Method

To overcome the safety concerns and inconvenience related to the utilization of highly corrosive acids, this method was developed. In a Babcock bottle, a combination of surfactants is mixed with a sample. The membrane of fat globule surrounding the emulsion droplets in milk is displaced by surfactants which causes them to coalesce and separate from each other. By the centrifugation of the sample, it allows the fat to go up in the graduated bottleneck, where we can determine its concentration.

7.4.3. Instrumental Methods

To determine the total lipid content present in food materials, there is a broad variety of instrumental methods available (Sahena et al., 2009; Pollardo et al., 2018). According to their physicochemical principles, these can be classified into following three different categories: (i) measurement of scattering of radiation, (ii) measurement of adsorption of radiation, and (iii) measurement of bulk physical properties. Each method has its pros and cons, and a series of foods to which it can be applied (Nishimura and Baker, 1987; Ryan et al., 2002a, b).

7.4.3.1. Measurement of Bulk Physical Characteristics/Properties

1. Ultrasonic velocity: The concentration of fat in food affect the speed at which an ultrasonic wave travels through a material. By

the measurement of ultrasonic velocity, the lipid content can be determined. This method is capable of nondestructive, rapid on-line measurements of lipid content.

2. *Electrical conductivity*: The conductivity of food decreases by the increase in lipid concentration because the electrical conductivity of lipids is way smaller than aqueous substances. Fat contents can be determined by the overall electrical conductivity of foods.
3. *Density:* There is a reduction in the density of food by the increase in fat content because the density of liquid oil is not as much as that of the majority of other food components. By measuring their density, the lipid content of foods can be determined.

7.4.3.2. Measurement of Adsorption of Radiation

1. *X-ray absorption:* The absorbance of X-ray decreases by the in crease in lipid concentration because lean meat more strongly absorbs X-rays as compared to fat. For the utilization of this phenomenon, to find out the lipid content in meat and meat products, commercial instruments have been developed.
2. *NMR:* The complete lipid concentration of foods is routinely determined by NMR spectroscopy. The area under a peak in an NMR chemical shift spectra corresponds to the lipid fraction and is thus used to determine the lipid content. By using instruments available commercially, lipid contents can be determined in a few seconds with no need for any sample preparation.
3. *Infrared*: This technique depends on the IR energy being absorbed due to molecular rotations or vibrations connected with fat molecules at a wavelength of 5.73 mm: the more the absorbance the greater the fat present. Once an appropriate calibration curve has been developed, IR is specifically useful for on-line and rapid analysis of lipid content.
4. *UV-visible:* Measurement of the absorbance of UV-visible radiation can be used to determine the concentration of certain lipids. The technique can be pretty labor-intensive and time-consuming because the lipid must be diluted and extracted in an appropriate solvent before analysis.

7.4.3.3. Measurement of Scattering of Radiation

1. *Ultrasonic scattering:* The absorption of ultrasound by an emulsion and ultrasonic velocity is related to the oil droplets concentration present. That's why ultrasonic scattering techniques can be used to determine the oil droplets concentration in emulsions of concentrated food.
2. *Light scattering:* As the emulsion turbidity is directly proportional to the oil droplets concentration present, the oil droplets concentration in food emulsions which are dilute can be calculated using a technique called light scattering.

Several instrumental methods have many advantages over the mentioned extraction techniques because they require less sample preparation, nondestructive, and calculations are often simple, precise, and rapid.

The techniques which depend on measuring the bulk physical properties of foods have a major drawback that between the total lipid content and the physical property of interest, a calibration curve must be prepared which might depend on the food matrix it is present in and the form of lipid present. Also, only foods with comparatively simple compositions can be analyzed by these techniques. It is complicated to disentangle the overall measurement contributed by fat from the contribution of the other components in food which have a varied concentration of different components.

7.5. DETERMINATION OF LIPID COMPOSITION

The previous section discussed the analytical methods to calculate the total concentration of lipids in foods, without taking into account the type of lipids present (Kolakowska, 2002; Gozen and Jesorka, 2012). Lipids are a highly diverse group of compounds having free fatty acids, di-, tri-, and monoacylglycerol, sterols phospholipids, vitamins A and D and carotenoids. Additionally, the majority of these sub-groups are chemically complex themselves. All triacylglycerols are esters of three fatty acid molecules and glycerol, despite that, the fatty acids can have different unsaturation, branching, chain lengths and positions on the glycerol molecule (Ang and Lyon, 1990; Ribeiro et al., 2009). A large number of different chemical species are present in lipid which consists of just triacylglycerols. Food scientists must know or can specify the total lipid concentration and the concentration of the different types of lipid molecules present (Nice and Catimel, 1999).

Below mentioned are some significant reasons for determining the type of lipids:

1. *Lipid oxidation.* The formation of aromas, undesirable off-flavors and potentially toxic compounds (e.g., cholesterol oxides) are caused by foods having high concentrations of unsaturated lipids which are specifically susceptible to lipid oxidation;
2. *Food Quality.* Mouthfeel, flavor, texture, and appearance, are the desirable physical characteristics of foods that depend on the type of lipids present;
3. *Legal.* Government regulations require the amounts of unsaturated, saturated, and polyunsaturated lipids, in addition to the amount of cholesterol, be written on food labels;
4. *Food Processing.* To adjust the optimum values of processing conditions e.g., temperatures, flow rates, etc, the manufacture of many foods depends on knowledge of the type of lipids present; and
5. *Adulteration.* By measuring the type of lipids present, and comparing them with the expected profile for an unadulterated sample, adulteration of oils and fats can be detected.

7.5.1. Sample Preparation

It is essential that for analysis the sample which is chosen must be representative of the original food's lipids, and also the properties are not varied before the analysis. The lipid must be available in a quite pure form for the analysis of lipids in food.

Therefore foods that are approximately completely lipids, e.g., lard or vegetable oil, olive oil usually require little sample preparation for analysis (Holt and North, 1985; Muraca et al., 1991). Yet, it is essential to purify and extract the lipid component before analysis for many other foods. Sometimes squeezing the oil out by simply applying pressure to food (e.g., some, seeds, nuts, and fish) can extract the lipids. However, most foods need more rigorous extraction methods, such as the nonsolvent or solvent extraction methods as mentioned in the previous section (White et al., 1979; Pernet and Tremblay, 2003).

Lipids are often melted (if already they are not liquid) once they have been separated and then centrifuged or filtered to remove any irrelevant matter. Additionally, to eliminate any residual moisture which may affect the analysis, they are dried. Like with any analytical procedure, avoiding the alteration in the properties of the component under analysis during the extraction process is important. Reduction in the oxidation of unsaturated lipids can be done by avoiding exposure to light and heat and flushing containers with nitrogen gas, or by adding antioxidants (Elias et al., 1981; De La Jara et al., 2003).

7.5.2. Separation and Analysis Through Chromatography

For analyzing and separating the properties of lipids, chromatography is amongst the most powerful analytical procedures, especially in combination with techniques used for the identification of the chemical structure of peaks, e.g., NMR or mass spectrometry (MS). In the chromatographic analysis, a mixture of the molecules to be separated is passed through a column containing a matrix which can selectively retard the molecular flow (Nagy and Nordby, 1970; Kirjavainen et al., 1996). Based on differing affinities of molecules in the mixture for the matrix in the column, they are separated. The more the affinity between the matrix and a specific molecule, the slower it passes through the column thus retarding its movement (Getz et al., 1970; Christie, 1985). Thus depending on the strength of interaction of different molecules with the matrix, they can be separated. After the column separated them, and they pass through an appropriate detector (e.g., flame ionization or fluorescence, UV-visible), the concentration of every molecule is determined. The complete profile of molecules that are present in a lipid can be determined by chromatography (Park and Goins, 1994; Weerheim and Ponec, 2001). This information can be used to: calculate the degree of lipid oxidation; the amounts of unsaturated, saturated, polyunsaturated cholesterol and fat; detect adulteration; the extent of radiation or heat damage; determine the antioxidants present. To analyze the lipids in foods different forms of chromatography are available, e.g., high-pressure liquid chromatography (HPLC), gas chromatography (GC), thin-layer chromatography (TLC).

7.5.2.1. Lipid fractions by TLC

TLC is mainly used to determine and separate the concentration of many types of lipid groups in foods, e.g., monoacylglycerols, diacylglycerols, triacylglycerols, phospholipids, cholesterol oxides, and cholesterol. A suitable absorbing material is coated on a TLC plate and placed in a suitable solvent. A minute amount of the sample of lipid is spotted on the TLC plate which has to be analyzed. With time, due to capillary forces, the solvent moves up the plate and depending on their affinity for the absorbing material, separates different lipid fractions. At the end of the separation, to make the spots visible, a dye is sprayed on the plate. Identification of the lipids present is possible by the comparison of distance covered by the spots with standards of known composition. Spots can be analyzed further using techniques, such as MS or NMR, GC.

This procedure allows quick analysis of lipids present in fatty foods and is inexpensive.

7.5.2.2. Fatty Acid Methyl Esters by GC

Free fatty acids and intact triacylglycerols are not easy to analyze using GC (requires that the lipids be able to vaporize in the instrument) because they are not very volatile. That's why lipids are often derivatized before analysis to enhance their volatility. Firstly triacylglycerols are saponified due to which they break them down to free fatty acids and glycerol, and are then methylated.

Triacylglycerol $\xrightarrow{CH_3OH,NaOH}$ methylated glycerol + Fatty acid methyl esters (FAMEs)

Methylation lessens the polarity and saponification lessens the molecular weight (MW), both of them increase the volatility of lipids. The quantity of different volatile FAMEs existing in the sample is then analyzed through GC. An appropriate organic solvent is injected into a GC injection chamber after the FAMES are dissolved. In the injection chamber, the sample is heated to volatilize the FAMES and after that by a heated carrier gas, they are carried into the separating column. As the FAMES pass through the column, based on differences in their polarities and MWs, they are divided into many peaks which are then quantified through an appropriate detector. Calculation of the concentration and type of fatty acids in the sample of original lipids is possible by the determination of the total fatty acid profile.

7.5.3. Chemical Techniques

A lot of chemical methods have been developed which provide information about the type of lipids found in edible oil and fats (Jones et al., 2004). These techniques only give information regarding the average properties of the components of lipid present, (e.g., amount of acids present or degree of unsaturation, the average MW) so as compare to chromatography these techniques are a lot cruder. Nevertheless, they are broadly used in research and industry because they do not need expensive apparatus and are simple to perform (Poon et al., 1981; Croxton et al., 2010).

7.5.3.1. Saponification Number

In a sample, a measure of the *average MW* of the triacylglycerols in the saponification number. By treatment with alkali, there is a breaking down of a neutral fat into fatty acids and glycerol; this process is called as Saponification:

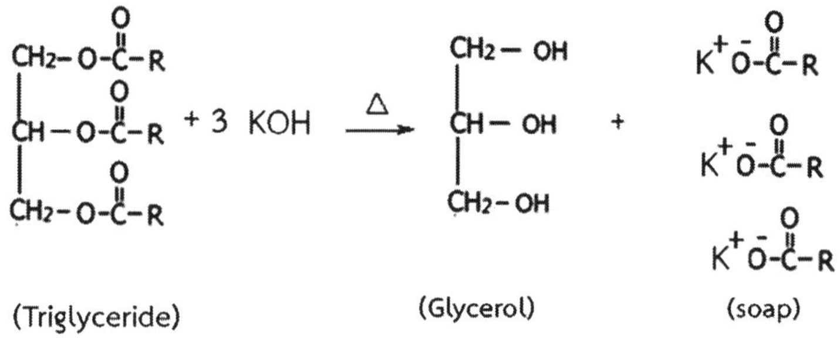

Figure 7.1: Saponification procedure.

Source: https://byjus.com/chemistry/saponification/.

The required mg of KOH to saponify 1 gram of fat is called a saponification number. Firstly, the lipid is extracted and after that, it is dissolved in an ethanol solution containing a well-known excess of KOH. So that the reaction goes to completion, it is then heated. Then by the addition of indicator and titrating the sample with HCl, the unreacted KOH is determined. The saponification number is calculated from the reacted amount of KOH and knowledge of the weight of the sample. The smaller the saponification number, the more will be the average MW of the triacylglycerols present.

7.5.3.2. Acid Value

The measure of *free acids* present in a specific amount of fat given is its acid value. The lipids are dissolved in an ethanol solution having an indicator after the extraction from a food sample. Then this solution is titrated with alkali (such as KOH) till the appearance of a pinkish color. The definition of the acid value is the mg of KOH required to neutralize the fatty acids which are present in 1 g of lipid. In the system, if other acid components are present, the acid value might be overestimated, e.g., acid phosphates or amino acids. The acid value is usually a good calculation of the breakdown of the triacylglycerols into free fatty acids, having a negative effect on the quality of numerous lipids.

7.5.4. Instrumental Methods

Information about lipid composition can also be provided by different instrumental methods. Nuclear magnetic resonance (NMR) spectroscopy is the most powerful amongst these. The concentration of particular types of chemical groups can be determined by the measurement of chemical shift spectra, which can be used to approximate the concentration of various types of lipids (Klinkner et al., 1997; Grazia Calvagno et al., 2007). Indirect information about the degree of unsaturation of oils and the average MW can be attained by measuring physical characteristics for e.g., refractive index (RI) or density. With increasing unsaturation and increasing chain length, the RI increases, however with decreasing unsaturation and increasing chain length, the density decreases. Therefore to monitor processes involving a change in the composition of oils, (such as hydrogenation, by which the degree of unsaturation decreases) measurements of density or the RI can be used (Archana et al., 2015).

7.6. METHODS USED FOR ANALYZING LIPID OXIDATION IN FOODS

Foods containing unsaturated lipids in high concentrations are specifically susceptible to the oxidation of lipids. It leads to the formation of potentially toxic and off-flavors compounds in foods, that's why it is amongst the main reason of food spoilage (St. Angelo et al., 1996; Xu et al., 2015). It is a highly complex process in which numerous reactions take place. These reactions give rise to many physical and chemical changes in Lipids:

reactants → primary products → secondary products

(unsaturated lipids and O_2) → (peroxides and conjugated dienes) → (aldehydes, ketones, hydrocarbons, alcohols).

There are a lot of methods developed by Food scientists to determine whether a specific lipid is susceptible to oxidation or not and to characterize the degree of lipid oxidation in foods.

7.6.1. Chromatography

Chromatography provides an elaborated profile of fatty acids and also the other molecules found in lipids which is why it is the strongest method to monitor the oxidation of lipids. By the measurement of changes in this profile with time, valuable information regarding the oxidation process of lipids can be obtained, especially by the identification of peaks using NMR or MS. By the use of chromatography, it is possible to monitor the specific reaction product formation (e.g., hydrocarbons or ketones, aldehydes) and the loss of reactants (e.g., unsaturated lipids). Such measurements might be made on components that are volatile in the headspace of a food, aqueous phase of food having water-soluble reaction products or on non-polar lipids that are extracted from the food (Shantha and Decker, 1994).

7.6.2. Oxygen Uptake

Lipid oxidation is dependent on the reaction between oxygen and unsaturated fatty acids. That's why as the reaction proceeds, by measuring the oxygen uptake by the sample, it is possible to observe the rate of its occurrence. Usually, the lipid is contained in a sealed container and the required amount of oxygen that must be put into the container to maintain the amount of oxygen in the head-space (HS) above the sample constant, is measured. The rate of lipid oxidation depends directly on the amount of oxygen that has to be put in the container. Therefore, this technique is an example of measuring the reduction in the concentration of reactants.

7.6.3. Conjugated Dienes

The non-conjugated double bonds found in natural unsaturated lipids (C=C-C-C=C) are transformed into conjugated double bonds (C=C-C=C), after the formation of peroxides. Conjugated trienes absorb ultraviolet (UV) radiation at 268 nm whereas conjugated dienes absorb strongly at 233 nm. Oxidation can be done by dissolving lipids in an appropriate organic solvent and by the use of a UV- visible spectrophotometer to measure alteration in its absorbance with time. In the later phases of oxidation of lipid, conjugated dienes (primary products) are fragmented into secondary products in which UV-visible light is not absorbed strongly which causes a decrease in absorbance. Therefore, this method is only useful to monitor the initial stages of lipid oxidation. This method is an example of measuring an increase in the quantity of primary reaction products.

7.6.4. Thiobarbituric Acid (TBA)

To determine the extent of lipid oxidation, this technique is amongst the most widely used. This method is used for calculating the concentration of comparatively polar secondary reaction products, such as aldehydes. The lipid which needs to be analyzed is dissolved in an appropriate non-polar solvent (present in a flask). To dissolve the polar secondary products in the sample, an aqueous solution of TBA reagent is added to the sample. After adding the reagent to the flask, the sample is shaken. As a result of shaking nonpolar solvent is separated from the aqueous phase. The aqueous phase is placed in a test-tube and is heated for twenty minutes in boiling water, due to which a pink color is produced. The intensity of this pink color is determined by the measurement of its absorbance at 540 nm, by the use of a UV-visible spectrophotometer. It is also related directly to the concentration of the TBA-reactive substance in the original sample. The basic source of color is the complex formation between malondialdehyde and TBA, although the TBA reagent can also be reacted with a few other secondary reaction products. That's why now this test is generally known as the TBA reactive substances (TBARS) method. It is an example of measuring an increase in the number of secondary reaction products.

7.7. CHARACTERIZATION OF PHYSICOCHEMICAL PROPERTIES

Because of the characteristic physicochemical properties of lipids such as appearance, texture, flavor, and mouthfeel, they are used in foods, in addition to their nutritional significance. During the other food preparation, they are also utilized as heat transfer agents, e.g., for frying. Therefore to characterize the lipid's physicochemical properties, food scientists must use analytical techniques (Westesen et al., 1997; Uner, 2006).

7.7.1. Solid Fat Content

Many physical and sensory properties, (such as mouthfeel, firmness, spreadability stability and processing) of lipids are influenced by its solid fat content (SFC). While characterizing lipids used in certain foods, e.g., butter, and margarine, food manufacturers mostly measure the SFC variation with temperature. The percentage of total lipid which is solid at a specific temperature is called as the SFC:

$$SFC = 100*M_{solid}/M_{total}$$

Where, M_{solid} = the mass of the lipid that is solid, and M_{total} = the total mass of the lipid in the food.

To measure the dependence of the SFC on temperature, many methods have been developed. The solid fat density is higher than liquid oil density, that's why when a fat crystallizes there is an increase in density and when it melts there is a decrease in its density. So it is possible to determine the SFC-temperature profile. For that purpose the density is measured over a range of temperatures:

$$SFC = \frac{(\rho - \rho_L)}{(\rho_S - \rho_L)} \times 100$$

Where, p = the density of lipid at a specific temperature, and p_S and p_L are the lipid densities if it were fully solid or fully liquid at the same temperature.

Dilatometry or bottles are usually used to determine the density.

Measurements are simpler and quicker to carry out through NMR that's why NMR based instrumental methods have replaced density measurements although the instrumentation is significantly more expensive. A radiofrequency pulse is applied to the sample after placing it into an NMR instrument. An NMR signal in the sample is induced in this. The decay rate of this signal depends on the lipid state whether it is liquid or solid. It is possible to differentiate between the two components because the signal decays way more rapidly from the solid fat than from the liquid oil (Calabresi et al., 1994; Chen et al., 2010).

Changes in SFC are also commonly monitored by the techniques based on differential scanning calorimetry (DSC). Heat is absorbed or evolved whenever lipid melts or crystallizes, this heat is measured by these techniques. It is possible to determine the SFC-temperature profile, the total lipid component involved in the transition and the melting point by making these measurements over a long range of temperatures.

7.7.2. Melting Point

In many cases, SFC over the entire range of temperatures is not essential to be known. Instead, the only information that is required is the temperature at which melting ends or starts. A pure triacylglycerol has only one melting point that happens at a particular temperature. However, food lipids have a broad range of different triacylglycerols, having their unique melting point for each one, and hence they melt over a broad variety of temperatures. Thus there are a lot of different ways to define the "melting point" of a food lipid, and each definition corresponds to a different amount of solid fat remaining. Amongst the most commonly used "melting points" some are:

1. *Clear point.* A minute amount of fat is kept in a capillary tube and after that, they are heated at a controlled rate. "Clear point" is the temperature at which the fat becomes transparent after it completely melts.
2. *Slip point.* A minute amount of fat is kept in a capillary tube and after that, they are heated at a controlled rate. The "slip point" is the temperature at which, due to the weight of fat it starts to move downwards.
3. *Wiley melting point.* A disc of fat is firstly suspended in a mixture of alcohol-water of similar density and after that, it is heated at a controlled rate. The "Wiley melting point" is the temperature at which the shape of disc changes to a sphere.

7.7.3. Cloud Point

The temperature at which crystallization of liquid oil begins is measured by its cloud point. A sample of fat is heated at such a temperature where all the crystals melt (for instance, 130°C). After that, the sample is cool down at a controlled rate and that temperature is determined at which the liquid just goes cloudy. It is the temperature at which crystals start to form and disperse light and this temperature is called the cloud point. Practically, it is important to have the type of oil, which when stored at 0°C, does not crystallize for long periods. A simple test is used to ascertain if a sample goes cloudy or not when kept for 5 hours at 0°C. It is used to determine the capacity of lipids to survive cold temperatures without the formation of crystals.

REFERENCE

1. Ai, Y., Hasjim, J., and Jane, J. L., (2013). Effects of lipids on enzymatic hydrolysis and physical properties of starch. *Carbohydrate Polymers*, *92*(1), 120–127.
2. Ang, C. Y. W., and Lyon, B. G., (1990). Evaluations of warmed-over flavor during chill storage of cooked broiler breast, thigh, and skin by chemical, instrumental, and sensory methods. *Journal of Food Science*, *55*(3), 644–648.
3. Archana, K. M., Ravi, R., and Anu-Appaiah, K. A., (2015). Correlation between ethanol stress and cellular fatty acid composition of alcohol-producing non-Saccharomyces in comparison with *Saccharomyces cerevisiae* by multivariate techniques. *Journal of Food Science and Technology*, *52*(10), 6770–6776.
4. Billing, B. H., Conlon, H. J., Hein, D. E., Schiff, L., Starr, C., and McGreevy, E., (1953). The value of needle biopsy in the chemical estimation of liver lipids in man. *The Journal of Clinical Investigation*, *32*(3), 214–225.
5. Byrdwell, W. C., (2001). Atmospheric pressure chemical ionization mass spectrometry for analysis of lipids. *Lipids*, *36*(4), 327–346.
6. Cai, S. S., and Syage, J. A., (2006). Comparison of atmospheric pressure photoionization, atmospheric pressure chemical ionization, and electrospray ionization mass spectrometry for analysis of lipids. *Analytical Chemistry*, *78*(4), 1191–1199.
7. Calabresi, L., Vecchio, G., Longhi, R., Gianazza, E., Palm, G., Wadensten, H., and Sejlitz, T., (1994). Molecular characterization of native and recombinant apolipoprotein A-I Milano dimer. The introduction of an interchain disulfide bridge remarkably alters the physicochemical properties of apolipoprotein AI. *Journal of Biological Chemistry*, *269*(51), 32168–32174.
8. Chen, C. C., Tsai, T. H., Huang, Z. R., and Fang, J. Y., (2010). Effects of lipophilic emulsifiers on the oral administration of lovastatin from nanostructured lipid carriers: physicochemical characterization and pharmacokinetics. *European Journal of Pharmaceutics and Biopharmaceutics*, *74*(3), 474–482.
9. Christie, W. W., (1985). Rapid separation and quantification of lipid classes by high performance liquid chromatography and mass (light-scattering) detection. *Journal of Lipid Research*, *26*(4), 507–512.

10. Croxton, R. S., Baron, M. G., Butler, D., Kent, T., and Sears, V. G., (2010). Variation in amino acid and lipid composition of latent fingerprints. *Forensic Science International, 199*(1–3), 93–102.
11. Dalzell, S. A., and Shelton, H. M., (1997). Methods of field preservation and selection of sample tissue for condensed tannin analysis in Leucaena species. *Animal Feed Science and Technology, 68*(3–4), 353–360.
12. De La Jara, A., Mendoza, H., Martel, A., Molina, C., Nordstron, L., De La Rosa, V., and Díaz, R., (2003). Flow cytometric determination of lipid content in a marine dinoflagellate, *Crypthecodinium cohnii. Journal of Applied Phycology, 15*(5), 433–438.
13. Dodds, E. D., McCoy, M. R., Geldenhuys, A., Rea, L. D., and Kennish, J. M., (2004). Microscale recovery of total lipids from fish tissue by accelerated solvent extraction. *Journal of the American Oil Chemists' Society, 81*(9), 835–840.
14. Donovan, J. M., Timofeyeva, N., and Carey, M. C., (1991). Influence of total lipid concentration, bile salt: lecithin ratio, and cholesterol content on inter-mixed micellar/vesicular (non-lecithin-associated) bile salt concentrations in model bile. *Journal of Lipid Research, 32*(9), 1501–1512.
15. Elias, P. M., Cooper, E. R., Korc, A., and Brown, B. E., (1981). Percutaneous transport in relation to stratum corneum structure and lipid composition. *Journal of Investigative Dermatology, 76*(4), 297–301.
16. Evershed, R. P., Turner-Walker, G., Hedges, R. E., Tuross, N., and Leyden, A., (1995). Preliminary results for the analysis of lipids in ancient bone. *Journal of Archaeological Science, 22*(2), 277–290.
17. Fennema, O., Donhowe, I. G., and Kester, J. J., (1994). Lipid type and location of the relative humidity gradient influence on the barrier properties of lipids to water vapor. In *Water in Foods* (Vol. 1, pp. 225–239). Pergamon.
18. Frings, C. S., and Dunn, R. T., (1970). A colorimetric method for determination of total serum lipids based on the sulfo-phospho-vanillin reaction. *American Journal of Clinical Pathology, 53*(1), 89–91.
19. Frings, C. S., Fendley, T. W., Dunn, R. T., and Queen, C. A., (1972). Improved determination of total serum lipids by the sulfo-phospho-vanillin reaction. *Clinical Chemistry, 18*(7), 673–674.
20. Galliard, T., (1968). Aspects of lipid metabolism in higher plants—I.

Identification and quantitative determination of the lipids in potato tubers. *Phytochemistry*, *7*(11), 1907–1914.

21. Gbogouri, G. A., Linder, M., Fanni, J., and Parmentier, M., (2006). Analysis of lipids extracted from salmon (Salmo salar) heads by commercial proteolytic enzymes. *European Journal of Lipid Science and Technology*, *108*(9), 766–775.
22. Getz, G. S., Jakovcic, S., Heywood, J., Frank, J., and Rabinowitz, M., (1970). A two-dimensional thin-layer chromatographic system for phospholipid separation the analysis of yeast phospholipids. *Biochimica et Biophysica Acta (BBA)-Lipids and Lipid Metabolism*, *218*(3), 441–452.
23. Go□zen, I., and Jesorka, A., (2012). Instrumental methods to characterize molecular phospholipid films on solid supports. *Analytical Chemistry*, *84*(2), 822–838.
24. Grazia Calvagno, M., Celia, C., Paolino, D., Cosco, D., Iannone, M., Castelli, F., and Fresta, M., (2007). Effects of lipid composition and preparation conditions on physical-chemical properties, technological parameters and in vitro biological activity of gemcitabine-loaded liposomes. *Current Drug Delivery*, *4*(1), 89–101.
25. Grima, E. M., González, M. J. I., and Giménez, A. G., (2013). Solvent extraction for microalgae lipids. In *Algae for Biofuels and Energy* (Vol. 1, pp. 187–205). Springer, Dordrecht.
26. Henson, J. M., Smith, P. H., and White, D. C., (1985). Examination of thermophilic methane-producing digesters by analysis of bacterial lipids. *Appl. Environ. Microbiol.*, *50*(6), 1428–1433.
27. Holt, W. V., and North, R. D., (1985). Determination of lipid composition and thermal phase transition temperature in an enriched plasma membrane fraction from ram spermatozoa. *Reproduction*, *73*(1), 285–294.
28. Hopmans, E. C., Schouten, S., Pancost, R. D., van der Meer, M. T., and Sinninghe, D. J. S., (2000). Analysis of intact tetraether lipids in archaeal cell material and sediments by high performance liquid chromatography/atmospheric pressure chemical ionization mass spectrometry. *Rapid Communications in Mass Spectrometry*, *14*(7), 585–589.
29. Jones, J. J., Stump, M. J., Fleming, R. C., Lay, J. O., and Wilkins, C. L., (2004). Strategies and data analysis techniques for lipid and

phospholipid chemistry elucidation by intact cell MALDI-FTMS. *Journal of the American Society for Mass Spectrometry*, *15*(11), 1665–1674.

30. Karabelnik, D., and Zbinden, G., (1975). Drug-induced foam cell reactions in rats, II. Chemical analysis of lipids stored in lungs and foam cells after treatment with chlorphentermine, 5-[p-(Fluoren-9-ylidenemethyl) phenyl]-2-piperidineethanol (RMI 10.393) and 1-Chloramitriptyline. *Hoppe-Seyler′ s Zeitschrift für Physiologische Chemie*, *356*(2), 1151–1160.
31. Kinsella, J. E., (1981). Functional properties of proteins: Possible relationships between structure and function in foams. *Food Chemistry*, *7*(4), 273–288.
32. Kirjavainen, M., Urtti, A., Jaaskelainen, I., Suhonen, T. M., Paronen, P., Valjakka-Koskela, R., and Mönkkonen, J., (1996). Interaction of liposomes with human skin in vitro—the influence of lipid composition and structure. *Biochimica et Biophysica Acta (BBA)-Lipids and Lipid Metabolism*, *1304*(3), 179–189.
33. Klinkner, A. M., Bugelski, P. J., Waites, C. R., Louden, C., Hart, T. K., and Kerns, W. D., (1997). A novel technique for mapping the lipid composition of atherosclerotic fatty streaks by en face fluorescence microscopy. *Journal of Histochemistry and Cytochemistry*, *45*(5), 743–753.
34. Koivusalo, M., Haimi, P., Heikinheimo, L., Kostiainen, R., and Somerharju, P., (2001). Quantitative determination of phospholipid compositions by ESI-MS: effects of acyl chain length, unsaturation, and lipid concentration on instrument response. *Journal of Lipid Research*, *42*(4), 663–672.
35. Kolakowska, A., (2002). Lipid oxidation in food systems. *Chemical and Functional Properties of Food Lipids, 1*, 133–160.
36. Liu, X., Wang, L., Zhang, J., Yin, J., and Liu, H., (2013). Global and local structure preservation for feature selection. *IEEE Transactions on Neural Networks and Learning Systems*, *25*(6), 1083–1095.
37. Long, R. D., and Abdelkader, E., (2011). Mixed-polarity azeotropic solvents for efficient extraction of lipids from Nannochloropsis microalgae. *Am. J. Biochem. Biotechnology*, *7*(2), 70–73.
38. McLemore, J. L., Beeley, P., Thorton, K., Morrisroe, K., Blackwell, W., and Dasgupta, A., (1998). Rapid automated determination of lipid

hydroperoxide concentrations and total antioxidant status of serum samples from patients infected with HIV: Elevated lipid hydroperoxide concentrations and depleted total antioxidant capacity of serum samples. *American Journal of Clinical Pathology*, *109*(3), 268–273.

39. Michalski, M. C., Genot, C., Gayet, C., Lopez, C., Fine, F., Joffre, F., and Steering committee of RMT LISTRAL. (2013). Multiscale structures of lipids in foods as parameters affecting fatty acid bioavailability and lipid metabolism. *Progress in Lipid Research*, *52*(4), 354–373.
40. Muraca, M., Vilei, M. T., Miconi, L., Petrin, P., Antoniutti, M., and Pedrazzoli, S., (1991). A simple method for the determination of lipid composition of human bile. *Journal of Lipid Research*, *32*(2), 371–374.
41. Nagy, S., and Nordby, H. E., (1970). Effects of storage conditions on the lipid composition of commercially prepared orange juice. *Journal of Agricultural and Food Chemistry*, *18*(4), 593–597.
42. Nice, E. C., and Catimel, B., (1999). Instrumental biosensors: New perspectives for the analysis of biomolecular interactions. *Bioessays*, *21*(4), 339–352.
43. Nichols, D. S., and Sanderson, K., (2003). The nomenclature, structure, and properties of food lipids. *Chemical and Functional Properties of Food Lipids*, *1*, 29–59.
44. Nickerson, C. J., and Lynch, L., (2001). The effect of farmland preservation programs on farmland prices. *American Journal of Agricultural Economics*, *83*(2), 341–351.
45. Nishimura, M., and Baker, E. W., (1987). Compositional similarities of non-solvent extractable fatty acids from recent marine sediments deposited in differing environments. *Geochimica et Cosmochimica Acta*, *51*(6), 1365–1378.
46. Noonan, D. S., (2007). Finding an impact of preservation policies: price effects of historic landmarks on attached homes in Chicago, 1990–1999. *Economic Development Quarterly*, *21*(1), 17–33.
47. Nourooz-Zadeh, J., Tajaddini-Sarmadi, J., Ling, K. E., and Wolff, S. P., (1996). Low-density lipoprotein is the major carrier of lipid hydroperoxides in plasma. Relevance to determination of total plasma lipid hydroperoxide concentrations. *Biochemical Journal*, *313*(3), 781–786.
48. Park, P. W., and Goins, R. E., (1994). In situ preparation of fatty acid methyl esters for analysis of fatty acid composition in foods. *Journal*

of Food Science, *59*(6), 1262–1266.

49. Passi, S., Rothschild-Boros, M. C., Fasella, P., Nazzaro-Porro, M., and Whitehouse, D., (1981). An application of high performance liquid chromatography to analysis of lipids in archaeological samples. *Journal of Lipid Research*, *22*(5), 778–784.
50. Pernet, F., and Tremblay, R., (2003). Effect of ultrasonication and grinding on the determination of lipid class content of microalgae harvested on filters. *Lipids*, *38*(11), 1191–1195.
51. Pfeiffer, S., and Varney, T. L., (2002). Quantifying histological and chemical preservation in archaeological bone. In *Biogeochemical Approaches to Paleodietary Analysis* (Vol.1, pp. 141–158). Springer, Boston, MA.
52. Pollardo, A. A., Lee, H. S., Lee, D., Kim, S., and Kim, J., (2018). Solvent effect on the enzymatic production of biodiesel from waste animal fat. *Journal of Cleaner Production*, *185*, 382–388.
53. Poon, R., Richards, J. M., and Clark, W. R., (1981). The relationship between plasma membrane lipid composition and physical-chemical properties II. Effect of phospholipid fatty acid modulation on plasma membrane physical properties and enzymatic activities. *Biochimica et Biophysica Acta (BBA)-Biomembranes*, *649*(1), 58–66.
54. Ribeiro, A. P. B., Basso, R. C., Grimaldi, R., Gioielli, L. A., and Gonçalves, L. A. G., (2009). Instrumental methods for the evaluation of interesterified fats. *Food Analytical Methods*, *2*(4), 282–302.
55. Ryan, K. J., Homco-Ryan, C. L., Jenson, J., Robbins, K. L., Prestat, C., and Brewer, M. S., (2002b). Lipid extraction process on texturized soy flour and wheat gluten protein-protein interactions in a dough matrix. *Cereal Chemistry*, *79*(3), 434–438.
56. Ryan, K. J., Homco-Ryan, C. L., Jenson, J., Robbins, K. L., Prestat, C., and Brewer, M. S., (2002a). Effect of lipid extraction process on performance of texturized soy flour added wheat bread. *Journal of Food Science*, *67*(6), 2385–2390.
57. Sahena, F., Zaidul, I. S. M., Jinap, S., Karim, A. A., Abbas, K. A., Norulaini, N. A. N., and Omar, A. K. M., (2009). Application of supercritical CO_2 in lipid extraction: A review. *Journal of Food Engineering*, *95*(2), 240–253.
58. Schafer, K., (1998). Accelerated solvent extraction of lipids for determining the fatty acid composition of biological material. *Analytica*

Chimica. Acta, *358*(1), 69–77.

59. Shantha, N. C., and Decker, E. A., (1994). Rapid, sensitive, iron-based spectrophotometric methods for determination of perorlride values of food lipids. *Journal of AOAC International*, *77*(2), 421–424.
60. Smedes, F., (1999). Determination of total lipid using non-chlorinated solvents. *Analyst*, *124*(11), 1711–1718.
61. St. Angelo, A. J., Vercellotti, J., Jacks, T., and Legendre, M., (1996). Lipid oxidation in foods. *Critical Reviews in Food Science and Nutrition*, *36*(3), 175–224.
62. Strasberg, S. M., Toth, J. L., Gallinger, S., and Harvey, P. R. C., (1990). High protein and total lipid concentration are associated with reduced metastability of bile in an early stage of cholesterol gallstone formation. *Gastroenterology*, *98*(3), 739–746.
63. Stübiger, G., and Belgacem, O., (2007). Analysis of lipids using 2,4,6-trihydroxy acetophenone as a matrix for MALDI mass spectrometry. *Analytical Chemistry*, *79*(8), 3206–3213.
64. Tarricone, K., Wagner, G., and Klein, R., (2015). Toward standardization of sample collection and preservation for the quality of results in biomonitoring with trees: A critical review. *Ecological Indicators*, *57*, 341–359.
65. Teo, C. L., and Idris, A., (2014). Enhancing the various solvent extraction method via microwave irradiation for extraction of lipids from marine microalgae in biodiesel production. *Bioresource Technology*, *171*, 477–481.
66. Uner, M., (2006). Preparation, characterization, and physico-chemical properties of solid lipid nanoparticles (SLN) and nanostructured lipid carriers (NLC): their benefits as colloidal drug carrier systems. *Die pharmazie-an International Journal of Pharmaceutical Sciences*, *61*(5), 375–386.
67. Weerheim, A., and Ponec, M., (2001). Determination of stratum corneum lipid profile by tape stripping in combination with high-performance thin-layer chromatography. *Archives of Dermatological Research*, *293*(4), 191–199.
68. Westesen, K., Bunjes, H., and Koch, M. H. J., (1997). Physicochemical characterization of lipid nanoparticles and evaluation of their drug loading capacity and sustained release potential. *Journal of Controlled Release*, *48*(2–3), 223–236.

69. White, D. C., Davis, W. M., Nickels, J. S., King, J. D., and Bobbie, R. J., (1979). Determination of the sedimentary microbial biomass by extractible lipid phosphate. *Oecologia*, *40*(1), 51–62.

70. Windham, W. R., Mertens, D. R., and Barton, F. E., (1989). Protocol for NIRS calibration: sample selection and equation development and validation. *Near-Infrared Reflectance Spectroscopy (NIRS): Analysis of Forage Quality*, *1*, 96–103.

71. Wolf, R. E., Morman, S. A., Hageman, P. L., Hoefen, T. M., and Plumlee, G. S., (2011). Simultaneous speciation of arsenic, selenium, and chromium: species stability, sample preservation, and analysis of ash and soil leachates. *Analytical and Bioanalytical Chemistry*, *401*(9), 2733.

72. Xu, J. L., Riccioli, C., and Sun, D. W., (2015). An overview on nondestructive spectroscopic techniques for lipid and lipid oxidation analysis in fish and fish products. *Comprehensive Reviews in Food Science and Food Safety*, *14*(4), 466–477.

73. Yang, L., and Paulson, A. T., (2000). Effects of lipids on mechanical and moisture barrier properties of edible gellan film. *Food Research International*, *33*(7), 571–578.

74. Young, G., Nippgen, F., Titterbrandt, S., and Cooney, M. J., (2010). Lipid extraction from biomass using co-solvent mixtures of ionic liquids and polar covalent molecules. *Separation and Purification Technology*, *72*(1), 118–121.

Chapter 8

Enzymes in Food Analysis

CONTENTS

8.1. INTRODUCTION

For times humans have been unintentionally tapping into the advantages of enzymes for the making of a range of foods before the discovery of enzymes and scientific know-how of its catalytic activity. It is not possible to consider food products such as wine, cheese, bread, and a large number of other alcoholic beverages without taking into account the role of enzymes in their manufacturing (Prodromidis and Karayannis, 2002; Cao et al., 2012). In the processing of a large number of food items, the distinctive features of enzymes such as their specificity and natural origins have been exploited since their discovery. The processes contain (1) transformation to fermentable sugars in brewing from dextrins; (2) use of pectinases for juice clarification; (3) saccharification and starch liquefaction by the use of amyloglucosidases and alpha-amylases; (4) by transformation to much sweeter keto-form of glucose using glucose isomerase; and (5) by the use of proteases in the brewing industry for chill-haze prevention just to mention a few. Cost-effective manufacturing of "customized" enzymes has been enabled by recent signs of progress in the area of protein engineering and molecular biology that have further widened the variety of food applications relating to enzymes. For instance, for oil degumming and increasing cheese yield, an enzyme used is phospholipases, lipases as a substitute to chemical processes for interesterification, carbohydrate oxidases used in the transplant production for lactose transformation to lacto-bionic for organ protection, and asparaginases in foods for acrylamide mitigation some of the evolving applications of enzymes in the food production process (Whitaker, 1990; Franck et al., 1995).

A very good example of challenges offered to the food analyst is the wide range of enzymes types that are being used in the production of food items, whose duty is to investigate the components or ingredients of various foods. Foods are essentially a multifaceted matrix of edible constituents, a number of these ingredients may be present in low amounts (e.g., antioxidants, micronutrients such as vitamins, flavor components) and others in high amounts (e.g., carbohydrates, macronutrients i.e., proteins, and lipids). These components can be analyzed using affordable and simple and more significantly, very sensitive and specific techniques to confirm that the "needle in the haystack" can be selected (recognized) and its size (quantity) determined. By technological progress in chromatography, mass spectrometry (MS), electrophoresis, etc. these challenges have been mostly achieved. For addressing these challenges, these techniques are versatile and preferably matched candidates because of their relative sensitivity and specificity (De Luca et al., 2005; Santiago-Felipe et al., 2014). The efficiency of these enzymes is influenced because of its protein nature by factors such as temperature, pH, and ionic strength catalytic milieu and the analytes. The scope of application of enzymes may be limited for this activity because they may compromise specificity, decrease the affinity of the enzyme for the analyte or substrate, denature of destabilizing the enzyme. By the use of immobilization methods, few of this limitation has been considered. Moreover, for the use of enzymes for this purpose, the significant progress in the field of biosensor technology gives a good opportunity by giving essential affordability, robustness, and versatility. Mainly, there are two components of the biosensor which is an analytical tool comprising of: (i) a transducer that conveys the resultant outcome of the bioreceptor-analyte contact to a measurable signal; (ii) a bioreceptor for food ingredient detection or selective analyte (Viswanathan et al., 2009).

These indications may be colorimetric, fluorometric or amperometric, piezoelectric, potentiometric. Moreover, commercial production has been made possible by the developments in the field of biotechnology, and hence accessibility, of several enzymes with the chosen features for use in such applications (Nunes et al., 1998; Monosik et al., 2012).

8.2. PROTEINS AND AMINO ACIDS

The main ingredient of the foods are proteins, and while the key interest in proteins is because of their functional and nutritional effects on foods, the key driver behind the development of highly sensitive and selective tools for the analysis and detection of proteins are the allergic reactions induced in persons by taking certain proteinaceous foods to ensure the conformation of food items to regulatory standards. For instance, lysozyme (muramidase) enzyme is approved for use derived from the hen egg white for protection of the value of ripened cheese as an antimicrobial agent. Though, the joint Food and Agricultural Organization/World Health Organization Codex Alimentarius Commission needs its existence in foods to be acknowledged on the label as an egg product as it is a potential allergen (Codex Standard I-1985). The main technique for analysis of these and other proteins has been enzyme-linked immunosorbent assay (ELISA) with detection limits ranging to 10 from 0.05 mg/kg (Schubert-Ulrich et al., 2009). Simple sample handling, specificity, high potential for automation and standardization and sensitivity are the features of ELISA for its widespread applications in these analyses. In various types of cheese Schneider et al. (2009) lately stated the analysis of lysozyme by ELISA. Their technique used the application of a peroxidase-labeled antibody that experiences a redox reaction with tetramethylbenzidine dihydrochloride to produce a substance that can be measured by the change in absorbance change at 450 nm. For the analysis of lysozyme wines, hen egg white, cheese, and other foods, many ELISA techniques have been reported (Rauch et al., 1990, Yoshida et al., 1991, Vidal et al., 2005; Kerkaert and De Meulenaer, 2007; Weber et al., 2007). Other proteins found in peanuts (Pomes et al., 2003), soy (You et al., 2008), hazelnuts (Holzhauser and Vieths, 1999), sesame (Husain et al., 2010), walnuts (Doi et al., 2008), etc. have also been analyzed using ELISA method. As many as 16 allergenic proteins have been found in soybeans of which the major culprit is β-conglycinin (Maruyama et al., 2001, Xiang et al., 2002). In the technique for β conglycinin analysis, You et al. (2008) used an epitope conforming to amino acid residues 78–84 of the protein to produce an immunogen by connecting it to chicken bovine serum albumin and ovalbumin as carrier proteins. By measuring absorbance at 492 nm, horseradish peroxidase was used as the enzyme for discovery (Rauch et al., 1990, 1995).

For a wide range of ingredients and food components, the major enzyme-based analytical tool is ELISA, and hence, a description of the fundamental principles is suitable here. The common principles relating to the use of competitive ELISA (which may be indirect or direct) and sandwiches are illustrated in Figures 8.1 and 8.2 for the analysis of protein. A capture antibody is initially immobilized onto a solid phase, in the direct sandwich ELISA, particular for the protein being analyzed. When analytes are present or introduced in the food material or system containing the protein of interest, they bind, particularly to the immobilized antibody. By introducing a second enzyme-based analyte-specific antibody into the food material, the discovery of such binding is accomplished by forming a sandwich. A second analyte-based antibody fixes to a secondary site to form the sandwich on the protein of interest (analyte), before the enzyme-based antibody binding in the indirect sandwich ELISA. This kind of ELISA mainly appropriates for large molecules like proteins because of the necessity for a secondary binding site or more than one epitope for particular binding. The binding reaction may be connected to a chromophore, in both types, with the protein concentration being proportional to the color intensity. Amperometric, piezoelectric, potentiometric, and electrochemical detection techniques have since been established and extensively used for these analyses.

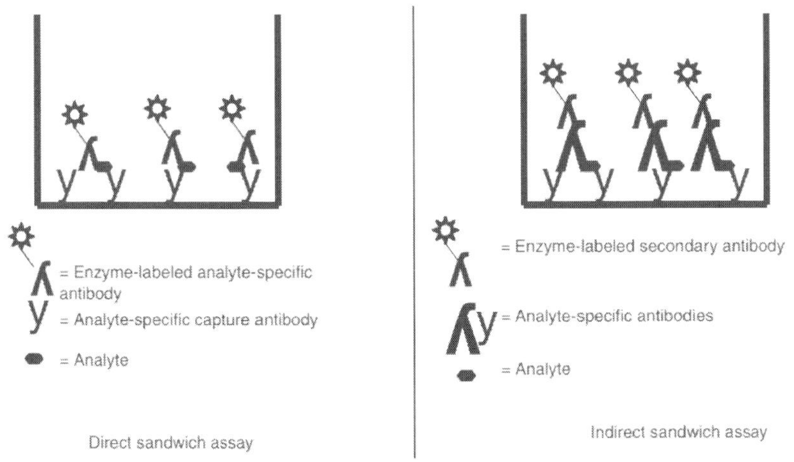

Figure 8.1: Sandwich ELISA with direct and indirect enzyme-labeled antibody detection.

Source: https://onlinelibrary.wiley.com/doi/pdf/10.1002/9781118308035. ch3.

For the allergen or protein analysis, other methods have also been established in later years, which uses enzyme-based antibodies for the detection. These comprise of the lateral flow examines, which are known as the dipstick tests, involving immobilized capture antibodies on a test band with analyte discovery by enzyme-based antibodies and immunochromatographic tests that involves the movement of immunoreactions along with a test band. Both these methods are normally quick, inexpensive, and portable and hence make them suitable for the online measurement. The reader is referred to the latest review by Schubert-Ulrich et al. (2009), for a thorough review of these methods and other techniques for allergen and protein analysis. For protein analysis, the development of enzyme-based biosensors has been advanced by the rising field of nanotechnology. By the use of carbon-nanotube-gold-nanoparticle hybrid connected to horseradish peroxidase, the latest report apprehended the spirit of this as a biosensor for determining the quantity of protein in serum (Cui et al., 2008). In this technique, the existence of hydrogen peroxide oxidation of phenylenediamine is catalyzed by the hybrid-bound peroxidase to produce diaminobenzene which generates an amperometric response because of its electroactive nature. There was no substantial difference between the two techniques when this method was compared with the ELISA.

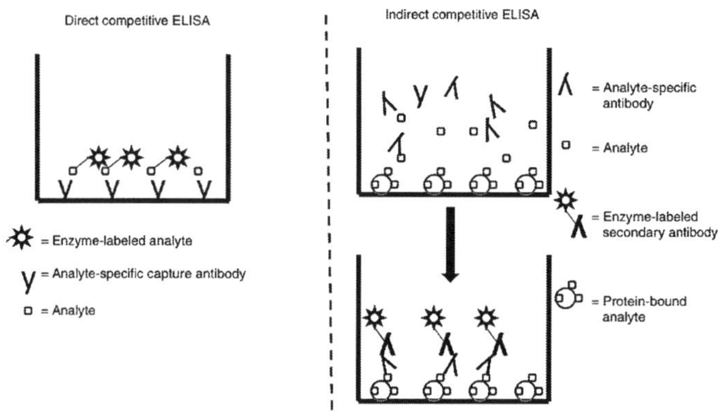

Figure 8.2: Competitive ELISA with direct and indirect enzyme-labeled antibody detection.

Source:-https://onlinelibrary.wiley.com/doi/pdf/10.1002/9781118308035.ch3.

Due to the wide impact of the amino acid content on food qualities and protein metabolism, the amino acid composition, as well as the free amino acid content of food proteins, establishes very significant food components. For instance, the savory flavor of umami is generally linked with the presence of glutamate, while bitterness in protein hydrolysates is caused by the high content of hydrophobic amino acids. Likewise, Maillard reactions are promoted by the provision of the free amino groups by the content of free amino acids, especially when these foods are treated with high-temperature conditions. Eventually, the color and flavor are introduced in the food products when these intermediates of the Maillard reaction undergo other reactions. In this era of the proteins-refreshed sports beverages, peptide, and free amino acid content of foods also give the athlete the "boost" required for the recovery from strong muscle activity during fitness exercise.

Application of the enzymes if involved in some analytical techniques for the measurement of amino acid content. For example, the utilization of amino acid decarboxylases to decarboxylate amino acids is one of the early techniques for amino acid analysis (AAA). By the use of glass electrodes having sodium (Na+) bicarbonate, the stoichiometric discharge of carbon (C) dioxide was then examined (Wiseman, 1981). Tryptophanase was utilized to catalyze the hydrolysis of the amino acid, in the analysis of tryptophan (Trp) to generate pyruvate. Coupling of the lactate dehydrogenase (LDH) to the pyruvate production converted NADH to NAD^+ and the equivalent alteration in absorbance at 340 nm was noted as the index of Trp content (Wiseman, 1981). Figure 8.3 shows the principle underlying the technique. In the aspartic aminotransferase examination of aspartic acid, an analogous technique involving combination to NADH has been exploited by conversion to oxaloacetate from aspartic acid in the existence of malate dehydrogenase (Wiseman, 1981). Several other enzyme-labeled techniques have been established for the AAA, with a substantial number of these also utilizing coupling to NADH-NAD^+ conversion.

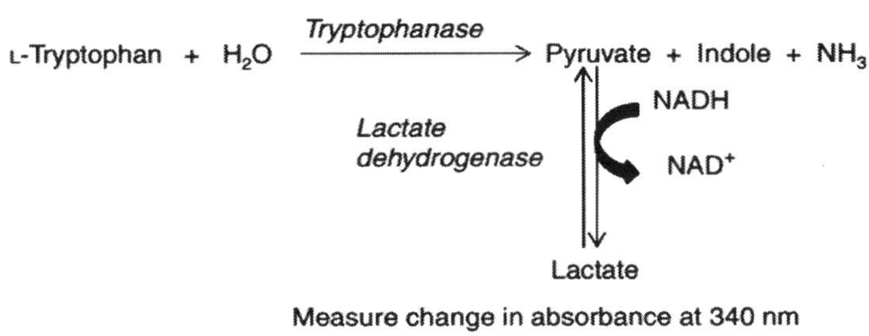

Figure 8.3: Principle underlying tryptophan analysis in foods using the coupled tryptophanase and lactate dehydrogenase (LDH) reaction.

Source: https://onlinelibrary.wiley.com/doi/pdf/10.1002/9781118308035.ch3.

8.3. SUGARS

By far, for the detection of glucose in foods the widely used and most predominant enzyme is Glucose oxidase. The formation of hydrogen peroxide and gluconic acid by the hydrolysis of glucose is the basis of the underlying principle. The former is then coupled to peroxidase catalysis, which results in the absorbance change by the oxidation of the colored precursor which is proportional to the glucose content (Wiseman, 1981). Other techniques are available which use enzyme combinations or complexes for glucose analysis. One of these techniques was based on the transfer of an acyl phosphate glucose-6-phosphate transferase catalyzed form glucose 6-phosphate from a donor molecule. Absorbance change was measured at 340 nm by coupling this reaction to NADP transformation to NADPH. For the determination of the sucrose (common sugar), glucose oxidase has also been used. The method involves the conversion of sucrose to α-d-glucose α-d-glucose and d-fructose by invertase-catalyzed conversion. After this conversion, the mutarotase-catalyzed isomerization β-d-glucose and α-d-glucose take place which is then oxidized to hydrogen peroxide and d-glucono-δ-lactone by glucose oxidase (Morkyavichene et al., 1984, Matsumoto et al., 1988, 2000). As explained earlier, the amount of peroxide released is proportional to the sucrose content resulting from the coupled reaction with peroxidase. Another monosaccharide that has a noteworthy interest in the food industry and has widespread use in various food products is Fructose. The indication of the level of adulteration, the ripening of food products such as the sweetness of several foods and honey is provided by the level of fructose in the foods. Based on highly specific pyrroloquinoline quinone (PQQ) enzyme, d-fructose dehydrogenase, there have been various analytical techniques (Paredes et al., 1997). The oxidation to 5-keto-d-fructose from d-fructose is catalyzed by the enzyme with the corresponding reduction of the covalently bound cofactor to $PQQH_2$ from PQQ. Coupling reoxidation to electrodes of the reduced cofactor generates an electric current in direct proportion to the quantity of fructose. Recently, Trivedi et al. (2009) reported an alternate method that involves the use of d-fructose dehydrogenase as an electron acceptor with ferricyanide. The principle underlying this technique is represented in Figure 8.4. With the reduction of ferricyanide, the oxidation of fructose to keto-fructose takes place by the enzyme. Subsequently, the reduced ferricyanide is re-oxidized, generating an electrical current proportional to the concentration of fructose.

For the analysis and detection of sugars and other food contaminants or components, Zeravik et al. (2009) is a recent review discussed the use of biosensors mentioned to as bioelectronic tongues. Ascorbic acid and glucose content of fruit juices have been examined by the use of one of these "tongues" involving metal catalysts (Pd, Au-Pd, and Pt) and glucose oxidase. By dropping the oxidation potential of the released hydrogen peroxide, the metal catalysts enhanced the sensitivity of the tongue as a result of the glucose oxidase action (Gutes et al., 2006). Alike simultaneous detection of sucrose, fructose, and glucose has been attained by the use of an immobilized enzyme reactor by amperometric flow injection analysis (Matsumoto et al., 1988, 2000). Zeravik et al. (2009) have explained the utilization of (a) laccase, peroxidase, and tyrosinase for phenolic compounds (b) PQQ-bound dehydrogenases for the measurement of the glucose (c) for sulfite content, sulfite oxidase, in the analysis of wine components. A complex polysaccharide, for analysis of starch, the underlying principles of saccharification and starch liquefaction are used. The starch is hydrolyzed by α-amylases, limiting dextrin and discharging maltose units. This is characterized by the commonly referred term in the starch industry as liquefaction also known as viscosity reduction. Additional treatment with amyloglucosidases and pullulanase outcomes in limit dextrin and hydrolysis of the maltose units, eventually generating glucose units which are then analyzed by some of the techniques discussed earlier.

Table 8.1: Enzymatic Methods for Food Component Analysis

Food Component	**Enzymes Involved**	**Commercial Kit**	**Reference**
Amino acids	Amino acid oxidase, amino acid decarboxylase		Wiseman, 1981, Mello and Kubota, 2002
Fructose	Fructose dehydrogenase, hexokinase, phosphoglucose isomerase, glucose-6-phosphate dehydrogenase	Sigma Aldrich, Inc. Megazyme International	
Raffinose	α-Galactosidase, β-galactose dehydrogenase, galactose mutarotase	Megazyme International	
Lysine *Carbohydrates*	Lysine oxidase		
Triglycerides	Lipase peroxidase, glycerol kinase, glycerol phosphate oxidase, glycerol phosphate oxidase, lipoprotein lipase, glycerol kinase	Cayman Chemical Co. Sigma Aldrich, Inc.	
Sucrose	Invertase, glucose-6-phosphate dehydrogenase, hexokinase	Megazyme International Sigma Aldrich, Inc.	
Tryptophan Aspartic acid Glutamic acid	Tryptophanase Glutamate dehydrogenase, Aspartate aminotransferase, diaphorase glutamate oxidase	Megazyme International	Prodromidis and Karayannis, 2002; Cock et al., 2009
Lactose	β-Galactosidase	Megazyme International	

Glucose	Glucose oxidase, glucose-6-phosphate dehydrogenase peroxidase, hexokinase	Cayman Chemical Co. Sigma Aldrich, Inc. Megazyme International	Wiseman, 1981; Mello and Kubota, 2002; Prodromidis and Karayannis, 2002; Cock et al., 2009
Pectin	Pectate lyase	Megazyme International	
Glycerol	Dehydrogenase, pyruvate kinase glycerol phosphate oxidase, glycerol-1-phosphate, glycerol dehydrogenase, lactate dehydrogenase, glycerol kinase, peroxidase	Megazyme International Cayman Chemical Co. Sigma Chemical Co.	Wiseman, 1981; Mello and Kubota, 2002; Prodromidis and Karayannis, 2002; Cock et al., 2009
Lactulose	β-galactosidase, fructose dehydrogenase		
Aspartame	Carboxypeptidase, carboxyl esterase, aspartase, glutamate oxidase, chymotrypsin, alcohol oxidase		
Lactate	Lactate dehydrogenase, transaminase		
Starch	Glucose-6-phosphate dehydrogenase, peroxidase, glucose oxidase, hexokinase, α-Amylase, amyloglucosidase,	Megazyme International Sigma Aldrich Company	
Catechol/ polyphenols	Tyrosinase, polyphenol oxidase		
Dietary fiber, dextran, and hemicellulose	Amyloglucosidase, α-Amylase, dextranase, hemicellulase protease	Sigma Aldrich, Inc.	

Phosphatidylcholine (lecithin)	Choline oxidase, phospholipase D, peroxidase	Cayman Chemical Co.	
Xylitol	Alkaline phosphatase (ELISA)		Sreenath and Venkatesh, 2010
Alcohol	Alcohol dehydrogenase	Megazyme International	Wiseman, 1981; Mello and Kubota, 2002; Cock et al., 2009
Cholesterol	Cholesterol oxidase, cholesterol esterase, peroxidase	Cayman Chemical Co.	
Ascorbic acid	Ascorbic acid oxidase	Megazyme International	
Citric acid	Citrate lyase		Wiseman, 1981; Mello and Kubota, 2002; Prodromidis and Karayannis, 2002; Cock et al., 2009
Oxalate	Oxalate oxidase		
Fatty acids	Lipases		

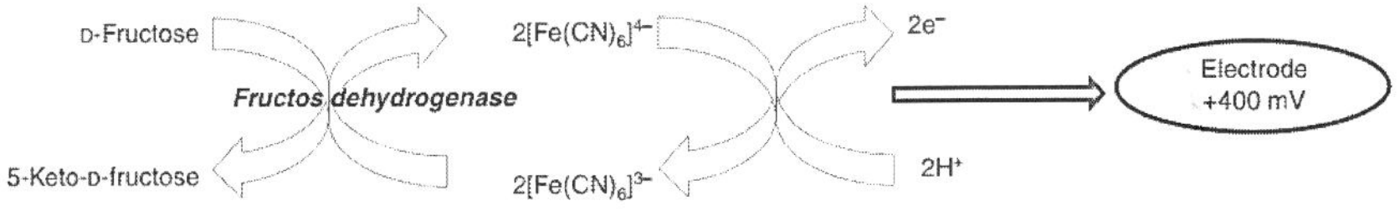

Figure 8.4: Schematic representation of the principle of fructose sensor based on fructose dehydrogenase.

Source: https://onlinelibrary.wiley.com/doi/pdf/10.1002/9781118308035.ch3.

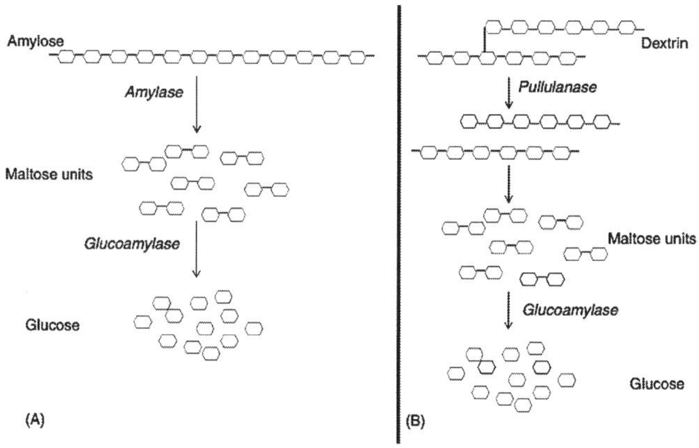

Figure 8.5: Schematic of enzymatic breakdown of starch (amylose and amylopectin) to glucose. (A) With amylose as substrate, amylase hydrolyzes the starch to maltose units, which are further simplified to glucose units. (B) With amylopectin substrate, pullulanase hydrolyzes the limit dextrin resulting from amylase activity. The products are further simplified by the amylase and glucoamylase to release glucose units.

Source: https://onlinelibrary.wiley.com/doi/pdf/10.1002/9781118308035.ch3.

8.4. ALCOHOL

In wine, beer, and other alcoholic drinks, alcohol (ethanol) content is a very significant quality attribute and is also essential for regulatory classification. In alcoholic beverages and foods, for analysis of alcohol content, the NAD-dependent alcohol dehydrogenase is extensively used. The oxidation of ethanol to acetaldehyde is catalyzed by the enzyme in an equilibrium reaction that also generates NADH (Mason and Dubowski, 1976; Jones and Schuberth, 1989). Acetaldehyde reaction with hydrazine or semicarbazide is used to proceed with the reaction in the forward direction. From NAD^+ to NADH conversion, the resultant absorbance change is proportional to alcohol content. For determining glycerol content in foods, similar dependent assays have been utilized. The principle underlying the technique for glycerol analysis has been shown in Figure 8.6 (Zimmerman and Maddrey, 1995; Cherpitel et al., 2003).

8.5. CHOLESTEROL

The excessive concentration in the level of cholesterol in foods has been produced by the utilization of cholesterol-rich foods and their connotation with hypertension, atherosclerosis, and high incidence of another cardiovascular disease. Cheese, beef, egg yolk, shrimps, poultry, and pork have a high level of cholesterol (Clarke et al., 1997; Brown et al., 1999). The basis of the initial enzymatic methods was colorimetric assays for the cholesterol analysis including primary de-esterification of cholesterol esters to release free cholesterol by cholesterol esterase, which is oxidized further to generate by cholesterol oxidase. In a peroxidase-catalyzed reaction, the peroxide formed is then paired to a colored or fluorescent reagent (Satoh et al., 1977, Hamada et al., 1980, Salè et al., 1984, Van Veldhoven et al., 2002). A viable kit (Boehringer's Monotest cholesterol or CHODPAP method) for the colorimetric assay, in which in the presence of the peroxide 4-aminoantipyrine and phenol are transformed to a red cyanogenimine dye released is obtainable. Control of the enzymes (cholesterol oxidase and cholesterol esterase) on carbon nanotubes and electrode surfaces have been generated by original and more sensitive electrochemical biosensors (Wisitsoraat et al., 2009). Likewise, for the detection of serum cholesterol which may also be appropriate to food examination, several enzyme-immobilized biosensors initially established. These contain agarose gel and activation with cyanogen bromide (Karube et al., 1982), pyrrole membrane joins with flow injection for hydrogen peroxide analysis (Wolfgang et al., 1993 Trialists, 2005, 2010) and immobilization on polyvinyl chloride with glutaraldehyde as a coupling agent (Hooda et al., 2009).

8.6. ANTIOXIDANTS

Because of the free-radical-induced off-flavors, the oxidation of lipids has been an old problem of the food industry and the involvement of these free radicals in cell harm and a number of pathogenetic conditions. In many food product formulations, numerous process modifications, as well as a combination of antioxidants control of such oxidations, has been accomplished. Though the ability of antioxidants to protect consumers against some of the potential health risks related to free radicals is considered as their additional responsibility to maintain the quality of food during storage and treatment. For antioxidant analysis an enzymatic assay recently developed involves oxidation of the yellowish compound syringaldazine by laccase with maximum absorbance at 530 nm to yield the purple-colored tetra methoxy azobismethylene quinone (TMAMQ). The creation of the purplish TMAMQ (Prasetyo et al., 2010) is blocked by the alteration of free radical intermediates in the presence of antioxidants. In a wide range of fruits and vegetables (e.g., lettuce, carrots, garlic, tomatoes, apples, kiwi, spinach,), beverages (e.g., coffee, green tea), and oils, this mechanism has been used to measure the level of antioxidant activity. However, for analysis of these ingredients and specific identification enzymes have been used in most cases, they have also been utilized in the withdrawal of some compounds for examination. Flavanoids, which usually occur in foods as the sulfated derivatives or glycosylated examined by enzymatic hydrolysis to the consistent desulfated and aglycones forms during their removal. For determination of naringenin and hesperetin in foods or β-glycosidases, β-glucuronidase and arylsulfatase are the most widely used in the removal process as found in the past report (Shinkaruk et al., 2010) extracted products were then exposed to peroxidase-linked ELIS.

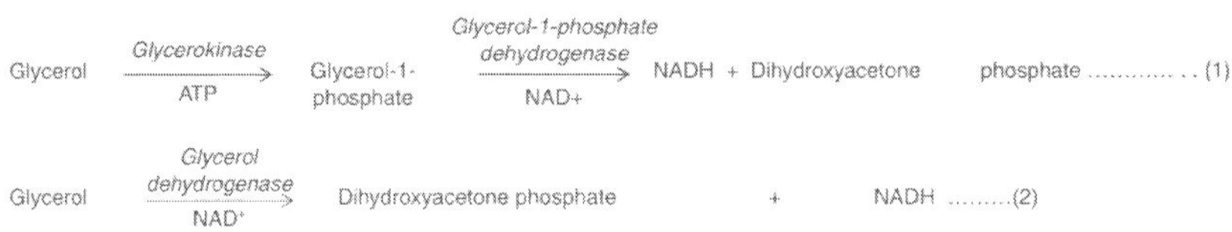

Figure 8.6: Glycerol kinase and glycerol dehydrogenase for glycerol analysis in foods.

Source: https://onlinelibrary.wiley.com/doi/pdf/10.1002/9781118308035.ch3.

REFERENCES

1. Alfonso, A., Vieytes, M. R., Yasumoto, T., and Botana, L. M., (2004). A rapid microplate fluorescence method to detect yessotoxins based on their capacity to activate phosphodiesterases. *Analytical Biochemistry, 326*(1), 93–99.
2. Brown, L., Rosner, B., Willett, W. W., and Sacks, F. M., (1999). Cholesterol-lowering effects of dietary fiber: a meta-analysis. *The American Journal of Clinical Nutrition, 69*(1), 30–42.
3. Cao, M., Li, Z., Wang, J., Ge, W., Yue, T., Li, R., and William, W. Y., (2012). Food related applications of magnetic iron oxide nanoparticles: Enzyme immobilization, protein purification, and food analysis. *Trends in Food Science and Technology, 27*(1), 47–56.
4. Cherpitel, C. J., Bond, J., Ye, Y., Borges, G., Macdonald, S., and Giesbrecht, N., (2003). A cross-national meta-analysis of alcohol and injury: Data from the emergency room collaborative alcohol analysis project (ERCAAP). *Addiction, 98*(9), 1277–1286.
5. Clarke, R., Frost, C., Collins, R., Appleby, P., and Peto, R., (1997). Dietary lipids and blood cholesterol: Quantitative meta-analysis of metabolic ward studies. *BMJ, 314*(7074), 112.
6. Cock, L. S., Arenas, A. M. Z., and Aponte, A. A., (2009). Use of enzymatic biosensors as quality indices: A synopsis of present and future trends in the food industry. *Chilean Journal of Agricultural Research, 69*(2), 270–280.
7. Cui, R., Huang, H., Yin, Z., Gao, D., and Zhu, J. J., (2008). Horseradish peroxidase-functionalized gold nanoparticle label for amplified immunoanalysis based on gold nanoparticles/carbon nanotubes hybrids modified biosensor. *Biosensors and Bioelectronics, 23*(11), 1666–1673.
8. De Luca, S., Florescu, M., Ghica, M. E., Lupu, A., Palleschi, G., Brett, C. M. A., and Compagnone, D., (2005). Carbon film electrodes for oxidase-based enzyme sensors in food analysis. *Talanta, 68*(2), 171–178.
9. Doi, H., Touhata, Y., Shibata, H., Sakai, S., Urisu, A., Akiyama, H., and Teshima, R., (2008). Reliable enzyme-linked immunosorbent assay for the determination of walnut proteins in processed foods. *Journal of Agricultural and Food Chemistry, 56*(17), 7625–7630.
10. Franck, M., Kolar, V., and Eremin, S. A., (1995). Enzyme immunoassays

for s-triazine herbicides and their application in environmental and food analysis. *Analytica Chimica Acta, 311*(3), 349–356.

11. Gutés, A., Ibanez, A. B., Del, V. M., and Cespedes, F., (2006). Automated SIA e-tongue employing a voltammetric biosensor array for the simultaneous determination of glucose and ascorbic acid. *Electroanalysis: An International Journal Devoted to Fundamental and Practical Aspects of Electroanalysis, 18*(1), 82–88.
12. Hamada, C., Iwasaki, M., Mibuchi, Y., Zaitsu, K., and Ohkura, Y., (1980). A sensitive enzymatic method for the fluorimetric assay of cholesterol in serum and high-density lipoprotein. *Chemical and Pharmaceutical Bulletin, 28*(10), 3131–3134.
13. Holzhauser, T., and Vieths, S., (1999). Indirect competitive ELISA for determination of traces of peanut (Arachis hypogaea L.) protein in complex food matrices. *Journal of Agricultural and Food Chemistry, 47*(2), 603–611.
14. Hooda, V., Gahlaut, A., Kumar, H., and Pundir, C. S., (2009). Biosensor based on enzyme coupled PVC reaction cell for electrochemical measurement of serum total cholesterol. *Sensors and Actuators B: Chemical, 136*(1), 235–241.
15. Jones, A. W., and Schuberth, J., (1989). Computer-aided headspace gas chromatography applied to blood-alcohol analysis: importance of online process control. *Journal of Forensic Science, 34*(5), 1116–1127.
16. Karube, I., Hara, K., Matsuoka, H., and Suzuki, S., (1982). Amperometric determination of total cholesterol in serum with use of immobilized cholesterol esterase and cholesterol oxidase. *Analytica Chimica Acta, 139*, 127–132.
17. Kerkaert, B., and De Meulenaer, B., (2007). Detection of hen's egg white lysozyme in food: Comparison between a sensitive HPLC and a commercial ELISA method. In: *234th American Chemical Society (ACS) National Meeting and Exposition 234, pp.* 1-44..
18. Mason, M. F., and Dubowski, K. M., (1976). Breath-alcohol analysis: Uses, methods, and some forensic problems—review and opinion. *Journal of Forensic Science, 21*(1), 9–41.
19. Matsumoto, K., Kamikado, H., Matsubara, H., and Osajima, Y., (1988). Simultaneous determination of glucose, fructose, and sucrose in mixtures by amperometric flow injection analysis with immobilized enzyme reactors. *Analytical Chemistry, 60*(2), 147–151.

20. Matsumoto, M., Maruyama, N., Watanabe, H., Kikuchihara, T., Kido, Y., and Tsuji, A., (2000). Residue analysis of clenbuterol in bovine and horse tissues and cow's milk by enzyme-linked immunosorbent assay. *Food Hygiene and Safety Science (Shokuhin Eiseigaku Zasshi)*, *41*(1), 48–53.

21. Mello, L. D., and Kubota, L. T., (2002). Review of the use of biosensors as analytical tools in the food and drink industries. *Food Chemistry*, *77*(2), 237–256.

22. Monosik, R., Stredansky, M., Tkac, J., and Sturdik, E., (2012). Application of enzyme biosensors in analysis of food and beverages. *Food Analytical Methods*, *5*(1), 40–53.

23. Morkyavichene, M. V., Dikchyuvene, A. A., Paulyukonis, A. B., and Kazlauskas, D. A., (1984). Choice of initial ratio of enzymes for immobilization in multienzyme systems [Penicillium vitale, mutarotase]. *Applied Biochemistry and Microbiology, 1, 1-44.*

24. Nunes, G. S., Toscano, I. A., and Barceló, D., (1998). Analysis of pesticides in food and environmental samples by enzyme-linked immunosorbent assays. *TrAC Trends in Analytical Chemistry*, *17*(2), 79–87.

25. Paredes, P. A., Parellada, J., Fernández, V. M., Katakis, I., and Domínguez, E., (1997). Amperometric mediated carbon paste biosensor based on D-fructose dehydrogenase for the determination of fructose in food analysis. *Biosensors and Bioelectronics*, *12*(12), 1233–1243.

26. Pomes, A., Helm, R. M., Bannon, G. A., Burks, A. W., Tsay, A., and Chapman, M. D., (2003). Monitoring peanut allergen in food products by measuring Ara h 1. *Journal of Allergy and Clinical Immunology*, *111*(3), 640–645.

27. Prasetyo, E. N., Kudanga, T., Steiner, W., Murkovic, M., Nyanhongo, G. S., and Guebitz, G. M., (2010). Laccase-generated tetra methoxy azobismethylene quinone (TMAMQ) as a tool for antioxidant activity measurement. *Food Chemistry*, *118*(2), 437–444.

28. Prodromidis, M. I., and Karayannis, M. I., (2002). Enzyme based amperometric biosensors for food analysis. *Electroanalysis: An International Journal Devoted to Fundamental and Practical Aspects of Electroanalysis*, *14*(4), 241–261.

29. Rauch, P., Hochel, I., and Kas, J. A. N., (1990). Sandwich enzyme immunoassay of hen egg lysozyme in foods. *Journal of Food Science*,

55(1), 103–105.

30. Rauch, P., Poplstein, M., Hochel, I., Fukal, L., Ferri, E., Abagnato, C. A., and Roda, A., (1995). Enhanced chemiluminescent sandwich enzyme immunoassay for hen egg lysozyme. *Journal of Bioluminescence and Chemiluminescence*, *10*(1), 35–40.
31. Salè, F. O., Marchesini, S., Fishman, P. H., and Berra, B., (1984). A sensitive enzymatic assay for determination of cholesterol in lipid extracts. *Analytical Biochemistry*, *142*(2), 347–350.
32. Santiago-Felipe, S., Tortajada-Genaro, L. A., Puchades, R., and Maquieira, A., (2014). Recombinase polymerase and enzyme-linked immunosorbent assay as a DNA amplification-detection strategy for food analysis. *Analytica Chimica. Acta*, *811*, 81–87.
33. Satoh, I., Karube, I., and Suzuki, S., (1977). Enzyme electrode for free cholesterol. *Biotechnology and Bioengineering*, *19*(7), 1095–1099.
34. Schneider, N., Weigel, I., Werkmeister, K., and Pischetsrieder, M., (2009). Development and validation of an enzyme-linked immunosorbent assay (ELISA) for quantification of lysozyme in cheese. *Journal of Agricultural and Food Chemistry*, *58*(1), 76–81.
35. Schubert-Ullrich, P., Rudolf, J., Ansari, P., Galler, B., Führer, M., Molinelli, A., and Baumgartner, S., (2009). Commercialized rapid immunoanalytical tests for determination of allergenic food proteins: an overview. *Analytical and Bioanalytical Chemistry*, *395*(1), 69–81.
36. Shinkaruk, S., Lamothe, V., Schmitter, J. M., Manach, C., Morand, C., Berard, A., and Bennetau-Pelissero, C., (2010). Development and validation of two new sensitive ELISAs for Hesperetin and Naringenin in biological fluids. *Food Chemistry*, *118*(2), 472–481.
37. Sreenath, K., and Venkatesh, Y. P., (2009). Quantification of xylitol in foods by an indirect competitive immunoassay. *Journal of Agricultural and Food Chemistry*, *58*(2), 1240–1246.
38. Trettnak, W., Lionti, I., and Mascini, M., (1993). Cholesterol biosensors prepared by electropolymerization of pyrrole. *Electroanalysis*, *5*(9–10), 753–763.
39. Trialists, C. T., (2005). Efficacy and safety of cholesterol-lowering treatment: prospective meta-analysis of data from 90 056 participants in 14 randomized trials of statins. *The Lancet*, *366*(9493), 1267–1278.
40. Trialists, C. T., (2010). Efficacy and safety of more intensive lowering of LDL cholesterol: a meta-analysis of data from 170 000 participants

in 26 randomized trials. *The Lancet*, *376*(9753), 1670–1681.

41. Van, V. P. P., Meyhi, E., and Mannaerts, G. P., (1998). Enzymatic quantitation of cholesterol esters in lipid extracts. *Analytical Biochemistry*, *258*(1), 152–155.
42. Vidal, M. L., Gautron, J., and Nys, Y., (2005). Development of an ELISA for quantifying lysozyme in hen egg white. *Journal of Agricultural and Food Chemistry*, *53*(7), 2379–2385.
43. Viswanathan, S., Radecka, H., and Radecki, J., (2009). Electrochemical biosensors for food analysis. *Monatshefte für Chemie-Chemical Monthly*, *140*(8), 891.
44. Weber, P., Steinhart, H., and Paschke, A., (2007). Investigation of the allergenic potential of wines fined with various proteinogenic fining agents by ELISA. *Journal of Agricultural and Food Chemistry*, *55*(8), 3127–3133.
45. Whitaker, J. R., (1990). New and future uses of enzymes in food processing. *Food Biotechnology*, *4*(2), 669–697.
46. Wiseman, A., (1981). Enzymes in analysis of foods. In *Enzymes and Food Processing* (Vol. 1, pp. 275–288). Springer, Dordrecht.
47. Wisitsoraat, A., Karuwan, C., Wong-ek, K., Phokharatkul, D., Sritongkham, P., and Tuantranont, A., (2009). High sensitivity electrochemical cholesterol sensor utilizing a vertically aligned carbon nanotube electrode with electropolymerized enzyme immobilization. *Sensors*, *9*(11), 8658–8668.
48. Xiang, P., Beardslee, T. A., Zeece, M. G., Markwell, J., and Sarath, G., (2002). Identification and analysis of a conserved immunoglobulin E-binding epitope in soybean G1a and G2a and peanut Arah 3 glycinins. *Archives of biochemistry and biophysics*, *408*(1), 51–57.
49. You, J., Li, D., Qiao, S., Wang, Z., He, P., Ou, D., and Dong, B., (2008). Development of a monoclonal antibody-based competitive ELISA for detection of β-conglycinin, an allergen from soybean. *Food Chemistry*, *106*(1), 352–360.
50. Zeravik, J., Hlavacek, A., Lacina, K., and Skládal, P., (2009). State of the art in the field of electronic and bioelectronic tongues-towards the analysis of wines. *Electroanalysis: An International Journal Devoted to Fundamental and Practical Aspects of Electroanalysis*, *21*(23), 2509–2520.
51. Zimmerman, H. J., and Maddrey, W. C., (1995). Acetaminophen

(paracetamol) hepatotoxicity with regular intake of alcohol: analysis of instances of therapeutic misadventure. *Hepatology*, *22*(3), 767–773.

INDEX

D

E

F

T

U

V

W